# Archaeological Chemistry

# About the Cover

The cover image is from the Toca do Serrote da Bastiana shelter. The photograph is courtesy of Niéde Guidon. The cover was designed by Leroy Corcoran of C&W Associates.

ACS SYMPOSIUM SERIES **831**

# Archaeological Chemistry

## Materials, Methods, and Meaning

**Kathryn A. Jakes,** Editor
*Ohio State University*

American Chemical Society, Washington, DC

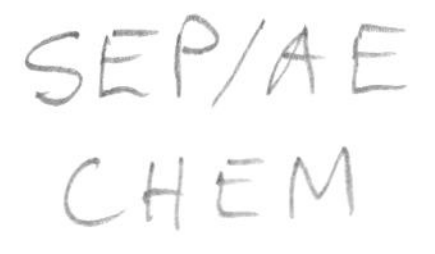

**Library of Congress Cataloging-in-Publication Data**

Archaeological chemistry : materials, methods, and meaning / Kathryn A. Jakes, editor.

p. cm.—(ACS symposium series ; 831)

Developed from a symposium sponsored by the Division of the History of Chemistry at the 222nd National Meeting of the American Chemical Society, Chicago, Illinois, August 26–30, 2001.

Includes bibliographical references and indexes.

ISBN 0–8412–3810–3

1. Archaeological chemistry. 2. Archaeology—Methodology.

I. Jakes, Kathryn A., 1949- II. American Chemical Society. Division of the History of Chemistry. III. American Chemical Society. Meeting (222nd : 2001 : Chicago, Ill..). IV. Series.

CC79.C5 A728 2002
930.1´028—dc21 2002025577

The paper used in this publication meets the minimum requirements of American National Standard for Information Sciences—Permanence of Paper for Printed Library Materials, ANSI Z39.48–1984.



Distributed by Oxford University Press



PRINTED IN THE UNITED STATES OF AMERICA

# Foreword

The ACS Symposium Series was first published in 1974 to provide a mechanism for publishing symposia quickly in book form. The purpose of the series is to publish timely, comprehensive books developed from ACS sponsored symposia based on current scientific research. Occasion-ally, books are developed from symposia sponsored by other organizations when the topic is of keen interest to the chemistry audience.

Before agreeing to publish a book, the proposed table of contents is reviewed for appropriate and comprehensive coverage and for interest to the audience. Some papers may be excluded to better focus the book; others may be added to provide comprehensiveness. When appropriate, overview or introductory chapters are added. Drafts of chapters are peer-reviewed prior to final acceptance or rejection, and manuscripts are prepared in camera-ready format.

As a rule, only original research papers and original review papers are included in the volumes. Verbatim reproductions of previously published papers are not accepted.

**ACS Books Department**

# Contents

## Indexes

# Preface

The chapters that comprise this volume provide a cross-section of current research in the area of chemistry applied to archaeological questions. The chapters were first presented as part of the 10th archaeological chemistry symposium, organized by the ACS Division of the History of Chemistry's Subdivision on Archaeological Chemistry, held in Chicago, Illinois, as part of the August 2001 national meeting of the American Chemical Society.

The chapters present examples of the use of analytical methods in the study of a variety of archaeological materials. They also discuss the inferences concerning human behavior that can be made based on the patterns observed in the analytical data. Viewed as a whole, the chapters encompass differing viewpoints concerning the definition of *nondestructive* testing and new developments in determining an object's age and composition. The volume provides a stimulating panoramic view of ongoing research useful to the archaeologist, anthropologist, and chemist. It also may spur interest in the novice reader. As we use chemistry to study the remains of the past, we uncover the traces of human behavior that are embodied in those remains.

**Kathryn A. Jakes**
Department of Consumer and Textile Sciences
College of Human Ecology
Ohio State University
1787 Neil Avenue
Columbus, OH 43210–1295

## Chapter 1

# Archaeological Chemistry: Materials, Methods, and Meaning

**Kathryn A. Jakes**

**Department of Consumer and Textile Sciences, Ohio State University, 1787 Neil Avenue, Columbus, OH 43210-1295**

The chapters that comprise this volume provide a cross section of current research conducted in the area of chemistry applied to archaeological questions. Each presents an example of the chemical analytical investigation of archaeological materials and the use of the analytical data in discerning patterns of human behavior. New developments in dating and compositional analysis are described. Differing views of the definition of "nondestructive" are presented.

Archaeology is an eclectic science, drawing from a variety of disciplines in the study of the materials that remain from past human activity. Applying geology, geography, history, anthropology, materials science, and chemistry, among others, we glean new knowledge of past lifeways. The chemist who undertakes an investigation of an object from an archaeological site can make a unique contribution to archaeology, often providing new information that is not apparent to the naked eye. We learn which foods people ate and the tools people used, also uncovering facts about how these materials were gathered locally or traded from afar. We are fascinated as we examine objects formed by early artisans, learning about how they were made and speculating about why they were made in a particular manner. We are connected to our predecessors on this planet as we learn what they knew, as we try to see what they saw, as we speculate about the things they valued. Spurred by the connection between us and the past, we probe the objects that embody clues to the past. Even as the world advances, we continue to be curious about those who have gone before us.

On first appraisal, the chapters that comprise this volume may appear to be a disparate group. The common thread that unites them is their use of chemical analytical methods to examine archaeological materials and their assessment of the information derived from these analyses to discern patterns of past human behavior. The chapters were first presented as part of the tenth archaeological chemistry symposium organized by the Division of History of Chemistry's Subdivision on Archaeological Chemistry and held as part of a national meeting of the American Chemical Society. Individually each provides an example of the current work being conducted in the area of chemistry applied to archaeological questions. Viewed as a whole they reflect the interdisciplinary nature of archaeological chemistry. Many different types of archaeological materials are discussed and many different chemical analytical methods are employed in the study of these materials. New developments in isotopic dating methods and in compositional analysis are described. The chapters also include some differing views of the definition of "nondestructive".

## Methods of Analysis

The methods described in the first chapters of this book reflect continued efforts by the archaeological chemist to glean data from artifacts while incurring the least possible sample destruction. They also present differing viewpoints concerning the definition of "nondestructive". Although there is general consensus among the authors that there should be no apparent destruction of the artifact in question, some differences can be seen in approach to this goal. Ciliberto (*1*) describes nondestructive analytical techniques as those which "allow analytical information to be obtained with no damage whatsoever to the sample or in some cases, the object in question. All visible alterations are avoided, and the object remains aesthetically unimpaired."

Archaeology is destructive by nature; some alteration is incurred when an object is excavated and brought into a new environment. Subsequent examination can incur damage to the recovered materials, causing fading of colors, for example. While Ciliberto (*1*) argues that there are "numerous fragments and leftover pieces" available for analysis of archeological objects, in some cases very little remains, and we strive to derive the most information possible from the smallest sample possible. New analytical methods help us achieve this goal, providing a vast amount of information from very small amounts of material. Two different approaches to the goal of nondestructive analysis are presented in these chapters. Either a very small sample is removed for subsequent examination or a very small amount of material is extracted from the bulk object. Is it best to remove a micro-sized sample, smaller than can be perceived by the naked eye? An assumption must be made in subsequent analyses that the results of the examination of small areas are representative of the object as a whole. On the other hand, if a very small amount of a component of the object in question can be removed from the entire object, the extract will provide a clue to the overall composition of the artifact. The extraction, however, will affect the chemistry of the entire object, even if only slightly. As the science of archaeological chemistry proceeds and new developments allow more information to be gleaned from ever smaller amounts of analyte, it will be important that any methods used in examination of the object be well documented.

In an innovative approach to obtaining the carbon for radioisotopic dating, Steelman and Rowe describe plasma chemical extraction of carbon from bulk organic objects. They tested the effects of the procedure on a peyote button, a clothing label, a sample of bone, and a sample of Third International Radiocarbon Intercomparison (TIRI) standard wood. The extraction process is "nondestructive in that it extracts only organic material" from the samples studied; the objects remain visibly unaffected by the treatment. The method is supported in the close match between the dates they obtained for the TIRI wood and charcoal samples and those obtained by other laboratories in the conventional manner. The plasma extraction procedure holds a great potential for dating organic archaeological artifacts. Although a sample may have to be removed from an object, the extraction procedure itself leaves no visible effect on the sample and additional tests can be performed.

In exploring the question of the earliest date for human presence in Brazil, Steelman and coworkers apply the plasma chemical procedure to extract carbon from both the pigments of a rock painting from Toca do Serrote de Bastiana and the accretions covering its surface. The accretions were found to contain both monohydrate of calcium oxalate and calcium carbonate. The radiocarbon age of the oxalate carbon was determined to be 2540 ± 60 B.P. while the radiocarbon age of carbon extracted from the pigment was determined to be 3730 ± 90 B.P. These ages are much more recent than the 30,000-40,000 B.P. age determined by electron spin resonance and thermoluminescence of the accretions, but are consistent with dates of other pictographs in the same shelter.

In another approach to extraction of a small amount of material from the bulk of an object, Reslewic and Burton extract lead from intact samples of majolica using ethylenediaminetetraacetic acid (EDTA). Lead isotope ratios of the EDTA extracts, determined by inductively coupled plasma mass spectrometry, are used to study the provenience of majolica. Their data agree with previous studies indicating that only a few production centers supplied New Spain with majolica in the 18$^{th}$ century. While more work needs to be done to determine the lead isotope ratio differences among geologic deposits and thereby identify distinct majolica groups, their EDTA extraction method holds a significant potential for the study of lead glazed pottery, and serves as a model for the concept of the use of chelating agents to extract components from archaeological materials with minimal effect on the objects.

Speakman and his colleagues describe the application of inductively coupled plasma mass spectrometry (ICP-MS) to the elemental analysis of obsidian, chert, pottery and painted and glazed surfaces. Only a very small area is affected in the laser ablation sampling, usually 1000 μm by 1000 μm by 30 μm. Their method is "virtually nondestructive" since the ablated area is not visible to the naked eye. Because ICP-MS has a lower detection limit than other common techniques, it is an ideal method for elemental analysis of materials with widely variable and yet localized composition.

## Materials Characterization

Several chapters in this volume present work conducted on the characterization of materials recovered from archaeological sites or of modern material prepared for comparative purposes. Chemical and physical attributes such as elemental composition, molecular structure, or isotopic ratios are determined and classification systems are developed that distinguish different members of a class. For example, Lambert and his coworkers examined modern and ancient plant resins employing solid-state carbon-13 nuclear magnetic resonance. They focus on the resins from Africa and the Americas, expanding on the classifications derived from results of infrared (IR) spectroscopy and gas chromatography-mass spectrometry (GC-MS). Thereby the composition of ancient artifacts can be used to determine their biological or geographic origin and inferences concerning exchange of resins or resinous objects also can be speculated. In a complementary paper, Lampert and her colleagues use GC and GC-MS to examine plant resins of Southeast Asia. They also identify characteristics of residues of archaeological resins coating the surfaces of ceramic potsherds, and are able to distinguish differences in chromatograms of resins from different plant genera. Radiocarbon dates of the resinous coatings of sherds from only one of the two sites under study agree with other known dates for these sites.

Similar work in characterization of materials is described in Chapter 8, in which Karklins et al. use neutron activation analysis to examine 290 glass beads and fragments from a glassmaking house in Amsterdam. The elemental composition, color, and pattern of the beads forms the basis by which beads found in sites in northeastern North America can be traced to their origins in Amsterdam. In Chapter 9, silk fibers from textiles recovered from the marine site of a wrecked ship are described by Srinivasan and Jakes. These seemingly fragile fibers are well preserved despite 130 years of submersion in the deep ocean. Not only can these textiles provide information about the people who traveled on the fated ship in 1857, they serve as models of the survival of fragile textiles in marine sites and thereby may encourage exploration and recovery.

Taking a different route to materials characterization, Hayes and Schurr use an experimental archaeological approach to develop a method for classification of archaeological maize and bone. They prepare comparative sets of maize and bone heated to different temperatures for differing periods of time and in conditions that include or exclude oxygen. Using Electron Spin Resonance (ESR), they are able to determine the maximum heating temperature and duration of heat treatment of carbonized plant remains and burned bone. In addition, because there are differences in the ESR parameters of maize recovered from middens and open pits, these parameters can be used to infer post deposit alterations in maize.

## Inference of Meaning

The strength of archaeological chemistry is revealed in the inferences that we make about past human behavior based on the study of the objects that remain. Patterns in the data derived from analysis of objects reflect patterns in their production and use. Therewith we infer their meaning, i.e., we determine those qualities that hold special significance. Lechtman describes the physical examination of an object that allows us "to discover specific technological styles which are renderings of appropriate technological behavior communicated through performance." (*2*). Not only can we determine how an object was made, we have a means to evaluate technological skills and choices of the object's manufacturer. Schiffer and Skibo (*3*) argue that some portion of the variability of an artifact is attributable to the person who made the artifact, and through understanding of the chain of events and relationships that led to the artifact's manufacture and use, inference of the causes for artifact variability can be drawn. Through systematic research we can "explain artifact designs rigorously and achieve an understanding of technology as a social process." In his unified theory of artifact design, Carr describes the "behavioral meanings" that can be inferred from "processes and constraints that are solely technological, to those that pertain to a society or community and its social segments of various decreasing scales, to those that operate at finer scales" (*4*). Behavioral meaning can extend from uncovering the method of manufacture of an artifact to discerning evidence of group social identity to identification of individual expressions of personal identity. While all the chapters in this volume provide examples of the use of analytical results applied to archaeological questions, the final chapters in particular provide examples of the application of the results derived from multiple techniques combined with contextual data to make inferences about human behavior.

In chapter 11, Tykot uses electron microprobe analysis to determine the major and minor elements in obsidian collected from sources throughout the Mediterranean region. Through statistical analysis, he is able to differentiate these sources and, thereby, is able to identify the sources of archaeological obsidian artifacts found in the region. Combining these data with contextual data, the relative frequency of particular obsidians in different sites can be used to infer that complex trade patterns existed in the Early Neolithic.

Combining multiple forms of analysis in their investigations of pottery manufacture in Northwest Mexico, Wells and Nelson infer aspects of the economic and political structure of the Malpaso Valley. Through analysis of clay deposits and sherds from sites in the La Quemada and Los Pilarillos settlements, two distinct chemical compositions were identified indicating that the potters used at least two distinctly different raw material sources. Distinctions between incised and engraved pots and range of firing temperature for the sherds reflects a change in production methods used by the potters. Because pottery decoration can be separated from the drying and firing processes leads to inferences of the possibility of craft specialization in an increasingly organized societal structure. In chapter 13, Reith also discusses examination of ceramic composition. X-ray fluorescence of ceramics of the Early Late Prehistoric period in eastern North America reveals no difference between Clemson Island ceramics, which are thought to be locally manufactured, and Owasco ceramics, which are thought to be foreign in manufacture and indicative of long distance interaction between groups. The similarity in composition of these artifacts implies that the same clay deposits were used in their manufacture and that the Owasco pots probably are not trade goods, but rather the distinctions in design and decoration of the two types of pots must reflect a local population's individual choices.

Tykot provides a comprehensive overview of research on diet of the Maya. Stable isotopic analysis of bone collagen and apatite yields information representative of the individual's average diet over a number of years while that of tooth enamel yields information representative of diet during the age of crown formation. From these data, he proposes that during the Classic Period, the Maya grew in dependence on maize. Not only is diet linked to age, sex, status, and local ecological factors, but it also can be linked to sociopolitical developments including increasing demands of the elite.

Energy dispersive x-ray fluorescence was used by Gaines and coworkers to determine the elemental composition of Chinese coins minted in the period of 1736-1796 and that of Annamese coins minted in the period 1880-1907. The consistency in composition of the two groups of coins, although minted 100 years apart, is indicative of a continued relationship between Annam and China. While Japanese modernization in the 19th century included new methods of metal refining and casting of coins, Annam maintained the traditional methods of coin manufacture.

## Summary

The chapters that comprise this volume present a cross section of current research in the area of chemistry applied to archaeological questions. Each chapter contributes to our knowledge of the past through study of material remains and discernment of patterns in the data that reflect patterns of human behavior. Viewed as a whole the chapters also provide a summary of recent developments in isotopic dating methods and in compositional analysis, as well as providing differing views of the definition of "nondestructive" testing.

## References


1. Ciliberto, E. In *Modern Analytical Methods in Art and Archaeology* ; Ciliberto, E; Spoto, G. Eds; Wiley-Interscience: New York, 2000; pp 1-10.
2. Lechtman, H In *Material Culture: Styles, Orgnazation and Dynamics of Technology*. Lechtman, H.; Merrill, R.S., Eds.; West Publishing,: St Paul, 1977; pp. 3-20.
3. Schiffer, M. B.; Skibo, J. M. *American Antiquity*. **1997**, *61*(1), 27-50.
4. Carr, C. *Style, Society and Person*. Carr, C; Neitzel, J.E., Eds.; Plenum: New York, 1995; pp. 171-258.

# Chapter 2

# Potential for Virtually Nondestructive Radiocarbon and Stable Carbon Isotopic Analyses on Perishable Archaeological Artifacts

**Karen L. Steelman and Marvin W. Rowe**

**Department of Chemistry, Texas A&M University, College Station, TX 77843-3255**

We present data using plasma-chemical extraction, originally developed to $^{14}C$ date rock paintings, for virtually 'non-destructive' $^{14}C$ dates from perishable artifacts. No visible alterations were observed in test samples after enough carbon was collected for both an AMS $^{14}C$ date and a stable isotope measurement. With many subsequent plasma reactions, minor changes were visible due to rigorous conditions that produced enough $CO_2$ for numerous radiocarbon dates. Radiocarbon dates can be corrected for natural mass fractionation with a $\delta^{13}C$ measurement. Mass fractionation during plasma reaction is negligible for almost all radiocarbon applications.

During the 1990s, the archaeological chemistry group at Texas A&M University developed a plasma-chemical procedure for extracting organic carbon from ancient rock paintings to $^{14}C$ date them (*1-16*). An advantage of this technique is its ability to selectively oxidize organic carbon in a sample, leaving accompanying carbonates and oxalates intact. This process is virtually non-destructive; it only removes organic matter from an ancient paint sample, with extracted $CO_2$ uncontaminated by carbonates, oxalates, or other mineral phases that are present. No visible changes in paint samples are discernable after plasma exposure, except that charcoal pigments slowly disappear during analysis. The sampling portion of the technique is destructive to pictograph images because a paint sample must first be removed from a rock shelter wall before laboratory analysis. But due to the small samples taken, negligible visible impact occurs. The reliability of $^{14}C$ dates for pictographs was show by dating pictographs with ages constrained by archaeological inference. Figure 1 summarizes previous results. Agreement is observed between archaeological inferences and $^{14}C$ dates (*8,9*), lending confidence to the validity of the method.

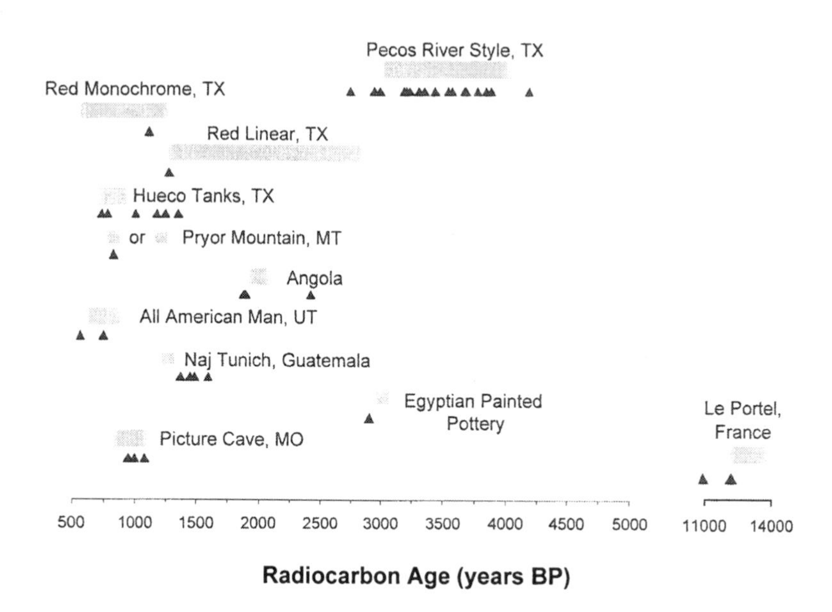


*Figure 1. Comparison of TAMU dates (triangles) with age ranges based on archaeological inferences (shaded bars). Agreement is observed in all cases.*

Pictographs represent a worst-case scenario for radiocarbon dating. If the pigment is not charcoal, the only organic material in the paint is most likely a binder or vehicle that was added to an inorganic pigment. After hundreds or thousands of years, the amount of organic material in the thin paint layer remaining on the wall is minuscule. Rowe discussed the attempts to chemically identify organic materials used in paints (*17*); little has been done.

Given typically only ~0.1 mg carbon is extracted from pictograph samples as $CO_2$, we realized that we should be able to obtain 'non-destructive' radiocarbon dates and stable isotopic carbon analyses on many perishable archaeological artifacts. Currently, for radiocarbon dating, subgram samples are removed from artifacts and destroyed during combustion extraction of $CO_2$. Plasma extraction does not require the removal of material from an object for destructive analysis. The characteristics of plasma extraction have great potential for the study of artifacts associated with burials, where the least intrusive techniques are especially critical. Museum artifacts such as basketry, textiles, cordage/twine, and paleobotanical samples would also benefit from this exciting application of the plasma extraction process.

Plasma extraction is 'non-destructive' in the sense that there is little or no observable physical or chemical change in the artifact; however, microscopic amounts of material would necessarily be removed by gentle surface oxidation and desiccation may be experienced. In some cases, simply placing an artifact under vacuum will cause serious deterioration in the structural integrity of the material. Each individual case would need to be discussed in advance by archaeologists, curators or conservators involved in the project. But even in the worst case, damage caused by plasma/vacuum processing cannot compare with complete combustion. Perhaps the technique should more accurately be termed minimally or virtually 'non-destructive' throughout, but for brevity it will be termed 'non-destructive' during the rest of this paper. The quotes are to be taken as a caution that the process is not absolutely and truly non-destructive.

Using argon and oxygen plasmas, we can extract a miniscule amount of carbon from the surface of a bulk artifact for analysis. A sample that initially weighed only as little as 10 mg need lose only a small fraction of its weight (AMS radiocarbon analysis can be done on ~0.1 mg carbon; stable isotope analysis requires even less, ~0.01 mg carbon). Archaeological samples are more often in the one-gram plus weight range; thus, a truly negligible fraction of the material need be removed, taken from the exposed surface of the sample and resulting in little or no observable erosion of the sample.

In the present study, we saw small, but visible, changes in a peyote button, a T-shirt label and Third International Radiocarbon Intercomparison (TIRI) standard wood samples (*5,6,8*) due to rigorous conditions and multiple plasma-chemical reactions used, with no visible changes in bone and charcoal. Because the charcoal samples with previously determined $^{14}C$ ages were on the order of a gram or more, there was no noticeable change in those samples. Charcoal pigment samples from rock paintings, however, do lose their black color when exposed to an oxygen plasma as the small amount simply reacts away.

Three questions are answered in this preliminary study. (a) Is the extent of mass fractionation occurring during plasma extractions negligible? (b) To what degree is the technique non-destructive? (c) Do the $^{14}C$ ages measured using the plasma technique agree with previously determined $^{14}C$ dates on samples? The answer to all these questions is a qualified yes. More specifically, the effect caused by ignoring mass fractionation results in an error of only ±32 $^{14}C$ years BP or less. Even that can be accurately corrected for as will be explained later. For all practical purposes, even with extreme plasma exposures used here, the plasma-chemical technique can be virtually 'non-destructive', making it a viable tool for radiocarbon dating rare archaeological artifacts. And, finally, agreement is observed between $^{14}C$ dates measured using plasma-chemical extraction compared to previously determined $^{14}C$ dates on aliquots of the same materials.

# Experimental Methods

## Plasma-chemical Extraction

Four different samples were studied: (a) a whole, modern peyote button weighing 1.490 g shown in Figure 2; (b) a modern label from a T-shirt weighing 0.352 g shown in Figure 3; (c) a sample of bone weighing several grams; and (d) charcoal samples weighing on the order of a gram each. In all cases, the plasma chamber was pre-cleaned of possible organic matter by running sequential oxygen (ultra high purity, 99.999%) plasmas until < 0.001 mg of carbon was released as $CO_2$. Then, the sample was placed into the chamber by unbolting a 2-3/4 inch stainless steel flange with a copper gasket under a positive pressure of ultra high purity argon (99.999%). Argon was flowed through the system and out the flange opening to retard back streaming of atmospheric gas into the system. After a sample was loaded in the chamber, several argon plasmas were run sequentially. Argon plasmas are non-reactive, and act as atomic 'sand-blasting', knocking loose adsorbed $CO_2$ (and possibly some of the more volatile organic materials) from the samples. Because the peyote and T-shirt label contained volatile compounds, presumably organic, we were not able to reduce the vacuum to correspond to <0.001 mg carbon equivalent for those samples as is usual in our procedure. Nonetheless, after performing several argon plasmas, we expect that adsorbed carbon dioxide was essentially removed and virtually all gaseous products frozen with liquid nitrogen (from the last argon plasma) were released volatile organic compounds. During the argon plasmas on the T-shirt label, cm-sized dark spots appeared on the chamber walls, confirming the presence of volatile organic compounds during this phase of operation. At that point, we began multiple, successive oxygen plasmas and collected $CO_2$ product after a sufficient period of time to remove enough carbon for replicate stable carbon isotope measurements and $^{14}C$ analysis, i.e., >0.160 mg of carbon for each oxygen plasma. The plasma power was arbitrarily selected as 100 watts, the power we have used for rock paintings. However, for 'non-destructive' extractions, lower powers may well be preferable. We subjected each sealed tube of $CO_2$ to carbon stable isotopic analysis to determine the extent of mass fractionation effects. For the archaeological samples, one aliquot was saved for AMS $^{14}C$ analysis. With future samples, multiple plasma oxidations would yield replicate $^{14}C$ dates and stable isotope measurements, allowing correction of a radiocarbon date for the actual $\delta^{13}C$ value of the $CO_2$ being dated.

## Radiocarbon analysis

For archaeological samples, $CO_2$ extracted with O-plasmas were sent to AMS laboratories for $^{14}C$ analyses. Radiocarbon dates were determined for three

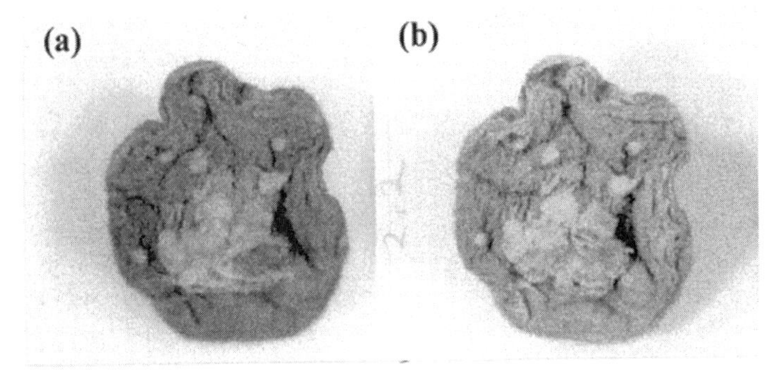


*Figure 2. Before (left-a) and after (right-b) photographs of the modern peyote button studied here. There was no noticeable change in the sample until the fourth oxygen plasma. With subsequent plasma exposures, some bleaching of the color was noted and is shown after 12 oxygen plasmas in the photograph on the right (b). Note the same white background is slightly more washed out (bleached looking) in the after photograph than in the before photograph. The peyote button is about 1 cm in diameter.*

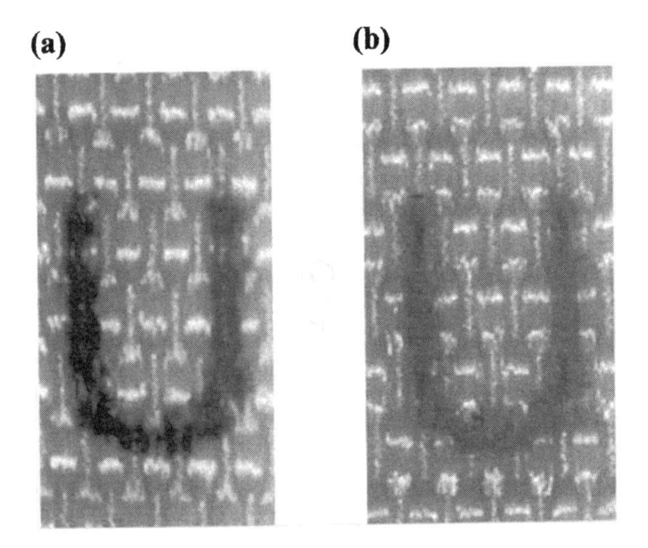


*Figure 3. Photographs before (left-a) and after (right-b) vacuum and four plasma exposures to the modern T-shirt label studied here. Some bleaching of the text color can be observed. Alterations were minor and no significant changes can be seen in the fabric. For scale, the height of the letter "U" is approximately 0.5 cm. No change was observed when the material was subjected to a plasma sufficient for a radiocarbon and a stable carbon isotopic analysis.*

Third International Radiocarbon Intercomparison (TIRI) standard wood samples (*5,6,8*) and four different archaeological charcoal samples (*2,3,18*) for comparison with $^{14}C$ dates previously determined at other laboratories using standard combustion methods. These measurements were utilized to demonstrate the viability of the plasma extraction technique for dating rock paintings from around the world with charcoal and inorganic pigments.

### Stable Isotope Measurements

Aliquots of $CO_2$ produced during O-plasmas were analyzed on three different stable isotope-ratio mass spectrometers over a period of five years. The peyote button and the T-shirt label were both run in the summer of 2001. These two samples were chosen to further establish that mass fractionation of carbon during the oxygen plasmas is a small or negligible effect and to test the 'non-destructive' nature of the technique on organic materials. For calibration to Peedee belemnite (PDB) primary standard, carbonate standard NBS-19 ($\delta^{13}C$ = 1.95‰) was used (i.e., Vienna PDB). $\delta^{13}C$ values are expressed relative to the V-PDB standard (*19*). Analytical precision was approximately ±0.06‰ for $\delta^{13}C$.

## Results

Table I shows our $^{14}C$ dates for the charcoal and TIRI wood samples that had been dated previously by other laboratories using standard pretreatment and combustion methods. Five replicate determinations of the fourth charcoal sample were conducted after using different pretreatments on aliquots of the same sample; possibly some variation in the radiocarbon dates could arise from that source. Nonetheless, statistical analysis indicated no discernable differences in our results and previously determined dates on the charcoal samples.

We have also subjected two modern samples to severe plasma treatments, far more vigorous than needed for $^{14}C$ dating and stable carbon isotopic analyses. These were (i) a piece of a T-shirt label, chosen because it had small writing that would make visual changes easy to see and (ii) a peyote button, selected because we had been asked to date an archaeological peyote button thought to be ~7-8,000 BP excavated from Shumla Cave in the Lower Pecos River region of Texas. The results of the $\delta^{13}C$ measurements on $CO_2$ produced from our plasma chemical extractions on five different samples are shown in Table II.

**Table I. Radiocarbon dates from the plasma extraction technique compared to previously determined radiocarbon ages for charcoal and TIRI wood**

| *Sample* | *TAMU $^{14}C$ date, BP* | *Reference* | *Previous $^{14}C$ date, BP* | *Reference* |
|---|---|---|---|---|
| Charcoal 1 | 3665 ± 65 | Russ et al. (*2*) | 3665 ± 60 | Tamers (*20*) |
| Charcoal 2 | 3650 ± 75 | Chaffee et al. (*3*) | 3655 ± 70 | Tamers (*20*) |
| Charcoal 3 | 3275 ± 75 | Chaffee et al. (*3*) | 3410 ± 75 | Tamers (*20*) |
| Charcoal 4a | 3350 ± 100 | Armitage et al. (*18*) | 3480 ± 80 | Turpin (*21*) |
| Charcoal 4b | 3500 ± 90 | Armitage et al. (*18*) | 3480 ± 80 | Turpin (*21*) |
| Charcoal 4c | 3770 ± 90 | Armitage et al. (*18*) | 3480 ± 80 | Turpin (*21*) |
| Charcoal 4d | 3540 ± 60 | Armitage et al. (*18*) | 3480 ± 80 | Turpin (*21*) |
| Charcoal 4e | 3510 ± 70 | Armitage et al. (*18*) | 3480 ± 80 | Turpin (*21*) |
| TIRI wood | 4730 ± 60 | Ilger et al. (*5*) | 4503 ± 6 | Gulliksen & Scott (*22*) |
| TIRI wood | 4530 ± 60 | Ilger et al. (*5*) | 4503 ± 6 | Gulliksen & Scott (*22*) |
| TIRI wood | 4530 ± 70 | Ilger et al. (*6*) Hyman & Rowe (*8*) | 4503 ± 6 | Gulliksen & Scott (*22*) |

## Discussion

### 'Non-destructive'?

One major objective of this paper was to conduct a preliminary investigation as to what degree the plasma-chemical extraction is non-destructive. The short conclusion is yes, the technique is virtually 'non-destructive', and in future studies damage to artifacts could be limited even further. For this preliminary examination, we selected two modern samples, a modern peyote button and a modern T-shirt label, as stringent cases to test the non-destructive nature of the process. Neither had been subjected to laboratory desiccation before introduction into the chamber. Thus, they were expected to contain a large fraction of their original organic components, making them likely to show combined effects of desiccation from the both vacuum and plasma treatments. Furthermore, the plasma extractions yielded far more carbon than was necessary for multiple $^{14}C$ dates and stable carbon isotopic analyses. Even so, the peyote button looked virtually unchanged to the human eye until 2.1 mg carbon had been extracted by four plasmas, theoretically enough for 20 $^{14}C$ dates, or 200 stable carbon isotopic analyses. Clearly, had the oxygen plasma been run for a much shorter amount of time, or perhaps at lower power, much less effect would have been seen. Figure 2 shows photographs of the peyote

**Table II. $\delta^{13}C$ Measurements on a modern peyote button and T-shirt label, and ancient bone and charcoal**

| *Sample* | *Time, min* | *Mass carbon, μg* | *$\delta^{13}C$, ‰* |
|---|---|---|---|
| **Peyote button** | | | |
| Run #1 | 15 | 960 | -15.86 |
| Run #2 | 5 | 410 | --- |
| Run #3 | 5 | 410 | -17.36 |
| Run #4[b] | 5 | 325 | -17.92 |
| Run #5 | 5 | 315 | -16.78 |
| Run #6[c] | 5 | 460 | -15.75 |
| Run #7 | 5 | 240 | --- |
| Run #8 | 5 | 275 | -17.44 |
| Run #9 | 5 | 325 | -17.01 |
| Run #10 | 5 | 375 | -16.77 |
| Run #11 | 5 | 285 | -16.95 |
| Run #12 | 5 | 185 | -17.01 |
| Average | | | -16.89±0.67(0.24)[a] |
| Bulk | | | -14.61 |
| **T-shirt label**[d] | | | |
| Run #1 | 10 | 295 | -25.90 |
| Run #2 | 10 | 285 | -26.57 |
| Run #3[e] | 15 | 380 | -24.55 |
| Run #4 | 13 | 420 | -25.94 |
| Average | | | -25.74±0.85(0.43)[a] |
| Bulk | | | -27.32 |
| **TIRI wood** | | | |
| Run #1 | 5 | 245 | --- |
| Run #2 | 5 | 270 | --- |
| Run #3 | 5 | 290 | --- |
| **Bone** | | | |
| Run #1 | 230 | 410 | -19.98 |
| Run #2 | 250 | 490 | -19.93 |
| Run #3 | 200 | 325 | -20.91 |
| Run #4 | 135 | 325 | -18.57 |
| Run #5 | 135 | 160 | -20.45 |
| Average | | | -19.97±0.88(0.39)[a] |
| **Charcoal** | | | |
| Run #1 | 15 | 175 | --- |
| Run #2 | 60 | 870 | -7.69 |
| Run #3 | 60 | 650 | -7.60 |
| Run #4 | 60 | 460 | --- |
| Run #5 | | | -7.61 |
| Average | | | -7.63±0.05(0.03)[a] |
| Bulk | | | -7.53±0.06 |

[a]the standard deviation is listed first, with the standard error of the mean shown in parentheses.

[b]Cracking due to desiccation was observed; subtle bleaching first noted.

[c]Bleaching was clearly observable.

[d]Dessication was observed after 1st argon plasma. During 2nd Ar plasma, dark spots were observed (deposition of volatile organics); spots started to disappear during 1st $O_2$ plasma; completely gone during 3rd $O_2$ plasma.

[e]Noticeable color change observed after 3rd $O_2$ plasma.

button before and after exposure to twelve oxygen plasmas. Although some bleaching was caused by the rigorous plasma treatments, it was relatively minor.

The modern T-shirt label also presented a challenging case because a written 'text' on the material could be watched carefully for alterations. Even for the face-up, exposed side, the writing appeared unchanged during the first two oxygen plasmas, in which enough carbon was extracted to provide five $^{14}C$ dates and eight separate stable carbon isotopic analyses. After the third oxygen plasma, a distinct fading of the writing was observed. The effect is demonstrated visually in Figure 3 after four total oxygen plasmas in which enough carbon was extracted for 12 radiocarbon and stable isotope analyses. The textile appears undamaged even at high magnifications.

Another more realistic, archaeological sample was a small piece of TIRI wood. Unlike the other two samples, TIRI wood is 4500 years old (*24*) and in excellent preservation. Three $^{14}C$ dates on this material were previously determined using plasma extraction (*5,6,8*). Although no visible changes were observed with the unaided eye during three oxygen plasma extractions, one section showed minor effects at high magnifications in an area where the wood was initially very thin (Figure 4a-b). Microscopic photographs illustrate the insignificance of the changes (Figure 4c-f).

Neither charcoal nor bone samples studied in our laboratory (*2,3,5,18*) have shown visible changes during plasma extraction. Extended plasma oxidation would, of course, completely remove all the carbonaceous material from charcoals given sufficient reaction time. Exposure to an ~1 gram sample long enough to provide five $^{14}C$ dates had no visible effect. From these studies, we are confident that carbon can be extracted virtually 'nondestructively', providing $CO_2$ for $^{14}C$ and stable isotopic measurements.

**Radiocarbon dates**

Another objective was to show that 'non-destructive' extraction of carbon permits reliable and accurate $^{14}C$ dates to be obtained. TIRI wood is a well dated specimen (4503±6 BP), being an international standard for comparing $^{14}C$ dates obtained at various laboratories (*22*). We have determined three dates using plasma chemical extraction. One of these was taken from the outer surface of the sample after it had been stored in a plastic bag for several years. The date on that sample was 4730±60 BP, older by 200 years than expected. Ilger et al. (*5*) attributed the difference to the possible inclusion of plasticizer from the plastic bag. Two additional TIRI wood samples were dated, this time ensuring that only interior samples were analyzed. The two additional dates, 4530±60 and 4530±70 BP (*5,6,8*), agreed very well with the accepted date of 4503±6 BP (*22*). Even the first age statistically agreed with the accepted date, although at the 2σ limit.

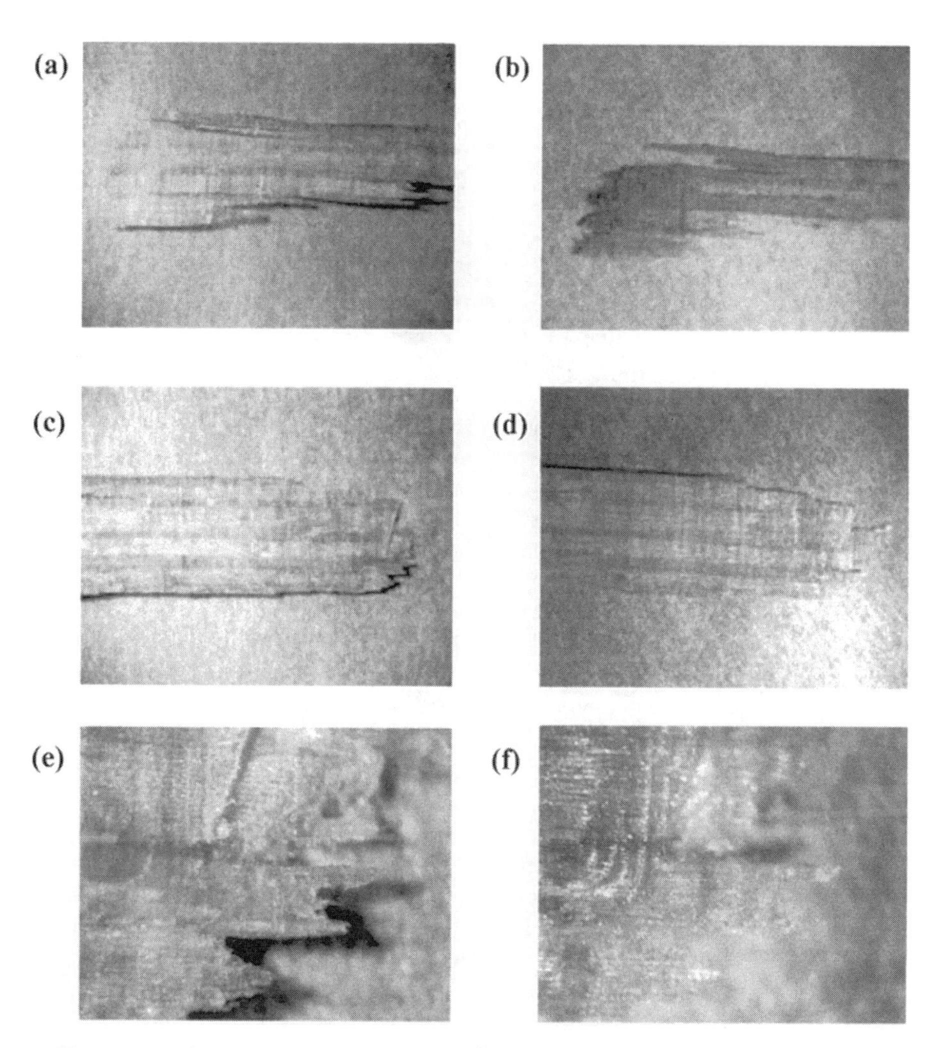


*Figure 4. Photographs before (left-a,c,e) and after (right-b,d,f) vacuum and plasma exposure to the 4500 year old TIRI wood studied here. For scale, the length of the wood piece shown in photographs a-d is ~2 cm and ~ 0.25 cm for e and f. Only minor damage is detected and none after extraction sufficient for one radiocarbon date and one stable carbon isotopic analysis.*

Statistical analysis shows no significant systematic variation in our results with plasma-chemical extraction and those determined at other laboratories using conventional and AMS $^{14}C$ analysis with combustion extraction. That includes the 'outlying' TIRI wood sample that may have been influenced by plasticizers (*5*) and effects introduced by different pretreatments of charcoal 4 (*18*). Our technique appears to produce accurate and viable $^{14}C$ dates.

## Mass fractionation

Finally, another objective of this work was to investigate whether mass fractionation was large enough to deleteriously affect radiocarbon dates using this plasma-chemical technique to extract organic carbon. Typically, for rock-art samples, we have observed almost complete reaction of organic material during the first oxygen plasma, run for approximately an hour. Therefore, mass fractionation is not an issue because there is nearly quantitative transfer of all the carbon in the sample to $CO_2$ for an AMS measurement. In the case of perishable archaeological materials, whole artifacts may be placed in the plasma chamber and only a miniscule amount of carbon extracted from the surface of the material. Given the lack of sample homogenization and quantitative transfer of carbon to $CO_2$, mass fractionation might be expected to occur to an extent that would seriously affect a $^{14}C$ date using this technique. There is a shift of only about 16 years with every per mil change from –25‰, the $\delta^{13}C$ value usually assumed by AMS radiocarbon laboratories in the absence of a direct $\delta^{13}C$ measurement for most samples (*23*). It is advantageous to measure the $\delta^{13}C$ value of extracted carbon dioxide so that any necessary corrections to account for *natural* mass fractionation can be made on a $^{14}C$ date. Mass fractionation occurs in nature through various biochemical reactions. The $\delta^{13}C$ value also gives some information as to the source of the material being dated.

To determine if there is significant mass fractionation occurring during exposure of the sample to oxygen plasmas, the $\delta^{13}C$ values of $CO_2$ produced from successive oxygen plasmas were compared, both for internal consistency and with $\delta^{13}C$ values of bulk material (when available). All $\delta^{13}C$ measurements here were determined by conventional stable isotope mass spectrometry. The means and sample standard deviations (s), along with the standard errors of the means ($s_m$ shown in parentheses) are reported in Table II. Q-tests at 95% confidence levels were performed on all $\delta^{13}C$ measurements taken here; no measurements were rejected.

The difference in the $\delta^{13}C$ values from the plasma and bulk measurements is most likely due to the very distinct modes of sampling carbon for the measurements. The bulk $\delta^{13}C$ measurements in Table II were performed on homogenized aliquots of the entire sample, as is traditionally done for stable isotope determinations. The $\delta^{13}C$ values for the $CO_2$ aliquots extracted with the plasma instead represent carbon removed from various surface components of a sample. It should not be surprising that a $\delta^{13}C$ value will vary depending upon the specific material being analyzed. For example, different plant parts are formed from a variety of biosynthetic pathways depending upon their chemical composition, i.e., cellulose, waxes, sugars, proteins, lipids, etc. (*24,25*). Significantly, measurements of $\delta^{13}C$ for $CO_2$ extracted with the plasmas from the 'homogenized' (ground) charcoal sample gave a 'plasma' precision indistinguishable from instrumental precision. That is, when the sample was homogeneous, no additional uncertainty was introduced by plasma extraction itself. Replicate $\delta^{13}C$ values on the two modern samples, however, indicated a real variation above the instrumental precision (±0.06 ‰). In all cases, the variation seen in the plasma extracted $\delta^{13}C$ values has a very minor effect on AMS radiocarbon analysis uncertainties. Furthermore, a correction for the actual $\delta^{13}C$ value of a given $^{14}C$ sample is possible as long as enough $CO_2$ is extracted to allow measurement of both the $^{14}C$ and the $\delta^{13}C$ at an AMS laboratory.

## Conclusions

Although preliminary, this report shows that 'non-destructive' $^{14}C$ and stable carbon isotopic measurements are feasible for some perishable archaeological artifacts. Mass fractionation appears unimportant, contributing no more than ~±32 year deflection on a $^{14}C$ age determination. In addition, mass fractionation can be corrected for by measuring the $\delta^{13}C$ in the same $CO_2$ aliquot used for AMS $^{14}C$ analysis. Visual inspection at high magnifications will continue to be used to confirm the 'non-destructive' nature of the technique. Consultation with archaeologists and conservators about the structural integrity of samples will also be made before future analyses. Samples will be archived to monitor their condition from any adverse effects due to plasma exposure. The most likely source of damage from plasma extraction would be from desiccation of the artifact under high vacuum in the sample chamber. Archaeological artifacts found in arid environments would expectedly be more suitable for this type of treatment. However, continued testing on modern samples that have not been desiccated will hopefully clarify the 'non-destructive' nature of the technique on samples found in more humid environments.

No analytical technique is entirely non-destructive. However, plasma-chemical carbon extraction is more attractive than combustion methods for some types of artifacts. While some deleterious effect may in the future be observed for artifacts that have undergone plasma extraction, removing a portion of the artifact for destructive combustion in many cases is not an option for a variety of reasons (i.e., ratio of artifact size to the amount that must be removed for analysis, information content of sample structure, rarity or uniqueness of object).

In fact, minor bleaching and desiccation, which would be of concern, were observed for test samples only after an unnecessarily excessive amount of carbon had been removed with the plasma. In the past, we have run oxygen plasmas at 100 watts power, yielding temperatures <150°C. We will investigate the efficacy of lower powers and hence lower temperatures for application on archaeological materials. It is expected that lower powers and shorter reaction times will minimize the impact on artifacts due to plasma exposure. While the term 'non-destructive' may be slightly misleading, this technique would allow artifacts to be $^{14}C$ dated that might not otherwise be accomplished.

In conclusion, the plasma technique has four advantages over ordinary combustion for $^{14}C$ dating (a) the plasma technique has been tentatively shown here to be 'non-destructive'; (b) the acid pretreatments normally used to rid the sample of carbonates and eliminate $CO_2$ adsorption are unnecessary with the plasma-chemical technique; (c) the background is very low; and (d) the plasma-chemical technique is unaffected by the presence of oxalates in the samples, unlike the traditional approach in some samples that are enriched in calcium oxalate (*15,26*).

## Acknowledgments

KLS was partially supported by a Regent's Fellowship from the Office of the Vice-Provost for Research (OVPR). Funding came from a Scholarly and Creative Activities Grant OVPR, with additional support from the Department of Chemistry and College of Science, Texas A&M University. Drs. Thomas Boutton, Ethan Grossman and Luis Cifuentes conducted stable carbon isotopic analyses. Martin Terry provided the peyote button. This paper was improved by thoughtful comments from an anonymous reviewer.

## References

1. Russ, J.; Hyman, M.; Shafer, H.; Rowe, M. W. *Nature* **1990**, *348*, 710-711.
2. Russ, J.; Hyman, M.; Rowe, M. W. *Radiocarbon* **1992**, *34*, 867-872.
3. Chaffee, S. D.; Hyman, M.; Rowe, M. W., In *New Light on Old Art: Recent Advances in Hunter-Gatherer Rock Art Research*, Whitley, D. S.; Loendorf, L. L. Eds.; Institute of Archaeology, University of California at Los Angeles, California, 1994, pp. 9-12.
4. Chaffee, S. D.; Hyman, M.; Rowe, M. W.; Coulam, N. J.; Schroedl, A.; Hogue, K. *American Antiquity* **1994**, *59*, 769-781.
5. Ilger, W. A.; Hyman, M.; Southon, J.; Rowe, M. W. *Radiocarbon*, **1995**, *37*, 299-310.

6. Ilger, W. A.; Hyman, M.; Southon, J.; Rowe, M. W. In *Archaeological Chemistry: Organic, Inorganic, and Biochemical Analysis*, Orna, M. V. Ed.; ACS Symposium Series 625: Washington, DC, 1995, pp 401-414.
7. Armitage, R. A.; Hyman, M.; Southon, J.; Barat, C.; Rowe, M. W. *Antiquity* **1997**, *71*, 715-719.
8. Hyman, M.; Rowe, M. W. *Techne* **1997**, *5*, 61-70.
9. Hyman, M.; Rowe, M. W. *Science Spectra* **1998**, *13*, 22-27.
10. Hyman, M.; Sutherland, K.; Rowe, M. W.; Armitage, R. A.; Southon, J. R. *Rock Art Research* **1999**, *16*, 75-88.
11. Pace, M. F. N.; Hyman, M.; Rowe, M. W.; Southon, J. R. In *American Indian Rock Art,* Bock, F. G. Ed.; American Rock Art Research Association, vol. 24, Tucson, AZ, 2000, pp 95-102.
12. Steelman, K; Waltman, E.; Rowe, M. W. *SBPN Scientific J.* **2000**, *4*, 90-97.
13. David, B.; Armitage, R. A.; Rowe, M. W.; Lawson, E. In *American Indian Rock Art,* Freers, S. M.; Woody, A., Eds.; American Rock Art Research Association, vol. 27, Tucson, AZ, 2001, pp 107-116.
14. Rowe, M. W. In *Handbook of Rock Art Research*, Whitley, D. S., Ed.; AltaMira Press, Walnut Creek, CA, 2001, pp 139-166.
15. Armitage, R. A.; Brady, J. E.; Cobb, A.; Southon, J. R.; Rowe, M. W. *American Antiquity* **2001**, *66*, 471-480.
16. Diaz-Granados, C.; Rowe, M. W.; Hyman, M.; Duncan, J. R.; Southon, J. R. *American Antiquity* **2001**, *66*, 481-492.
17. Rowe, M. W. In *Handbook of Rock Art Research*, Whitley, D. S., Ed.; AltaMira Press, Walnut Creek, CA, 2001, pp 190-220.
18. Armitage, R. A. Radiocarbon Dating of Charcoal-pigmented Rock Paintings; Ph.D. Dissertation, Texas A&M University; UMI Dissertation Services: Ann Arbor, MI, 1998.
19. Coplen, T. B. In *Reference and Intercomparison Materials for Stable Isotopes of Light Elements*. Vienna: IAEA, 1995, pp. 31-34.
20. Tamers, M. Personal communication, 1990.
21. Turpin, S. A. In *Papers on Lower Pecos Prehistory*, Turpin, S.A., Ed.; Studies in Archeology 8, Texas Archeological Laboratory, The University of Texas, Austin, Texas, pp.1-49.
22. Gulliksen, S.; Scott, M. *Radiocarbon* **1995**, 37, 820-821.
23. Bowman, S. *Interpreting the Past: Radiocarbon Dating*, University of California Press, Berkley, California, 1990; p. 21.
24. Ehleringer, J. R. In *Carbon Isotope Techniques*, D. C. Coleman, B. Fry, Eds.; Academic Press, Inc., San Diego, 1991, pp. 187-200.
25. O'Leary, M. H. *Phytochemistry* **1981**, *20*, 553-567.
26. Hedges, R.E.M.; Ramsey, C.B.; Van Klinken, G.J.; Pettitt, P.B.; Nielsen-March, C.; Etchegoyen, A.; Fernandez Niello, J.O.; Boschin M.T.; Llamazares A.M. *Radiocarbon* **1998**, *40*, 35-44.

## Chapter 3

# Accelerator Mass Spectrometry Radiocarbon Ages of an Oxalate Accretion and Rock Paintings at Toca do Serrote da Bastiana, Brazil

**Karen L. Steelman[1], Richard Rickman[1], Marvin W. Rowe[1], Thomas W. Boutton[2], Jon Russ[3], and Niéde Guidon[4]**

**Departments of [1]Chemistry and [2]Rangeland Ecology and Management, Texas A&M University, College Station, TX 77843**
**[3]Department of Chemistry, Arkansas State University, State University, AR 72467**
**[4]Fundação Museu do Homem Americano (FUMDHAM), Saõ Raimundo Nonato, Piaui, Brazil**

At the Toca do Serrote da Bastiana rock shelter in Brazil, a red, iron ochre pictograph of an anthropomorphic figure had become coated with a 'calcite' accretion over time. Using X-ray diffraction and Fourier transform infrared spectroscopy, we determined that the accretion also contains whewellite, the monohydrate of calcium oxalate, in addition to calcium carbonate. An AMS radiocarbon date on the oxalate formed from contemporaneous carbon yielded a minimum date of

2490 ± 30 years BP for the painting. An AMS radiocarbon age of 3730 ± 90 years BP on organic material in the underlying paint layer is consistent with the oxalate result and four direct radiocarbon dates on organic matter extracted from other rock paintings in the same shelter. Our results strongly disagree with a 30,000-40,000 year old age obtained by electron spin resonance and thermoluminescence dating of the accretion.

Controversy has reigned for years about the antiquity of an early human presence in Brazil (*1-4*). Work under the supervision of Niéde Guidon and the Fundação Museu do Homen Americano suggests that humankind was present in Brazil as early as 50,000 years ago, tens of thousands of years before humans were thought to have migrated into the Americas (*5, 6*). Whereas charcoal and quartzite flakes found at the Pedra Furada site may possibly be linked to natural phenomena *(7-9*), dates on rock art have the advantage that they are most certainly associated with a human presence (*10*). Currently, the Fundação Museu do Homen Americano has initiated research to locate proof of natural fire vestiges in the area surrounding Pedra Furada, but none has been found (*11*). Dates obtained by electron spin resonance (ESR) and thermoluminescence (TL) for the crystallization age of a 'calcite' accretion layer deposited over an ~20 cm tall red anthropomorphic pictograph in the Toca do Serrote da Bastiana shelter located in the state of Piaui further support the early arrival of humans in Brazil (Figure 1). The ESR and TL results suggest that the accretion was formed 30,000-40,000 years ago (*12*). Considering that the pictograph must have been produced prior to the 'calcite' deposition on top of it, this fascinating result would place the pictograph among the oldest dated rock paintings in the world.

In order to verify these results, we tested for calcium oxalate in the accretion layer covering the pictograph in question at the Toca do Serrote da Bastiana shelter. Calcium oxalate crusts are formed from contemporaneous carbon and are ubiquitous as accretions on limestone surfaces (*13-23*). Lichen and other bio-agents are thought to produce oxalate-rich rock crusts, although some nonbiological mechanisms have also been proposed (*13-15, 17, 21-24*). Oxalate crusts have been radiocarbon dated in order to obtain maximum and minimum ages for pictographs (*22, 23, 25-28*). We radiocarbon dated the oxalate portion of the accretion layer covering the red anthropomorphic pictograph in the Toca do Serrote da Bastiana shelter giving an age of 2490 ± 30 years. If the 30,000-40,000 TL/ESR age had been confirmed by independent means, the antiquity of this painted figure would provide strong support that a human presence was in Brazil far earlier than generally accepted. Unfortunately, that was not the case.

*Figure 1. The red anthropomorphic figure on the left was the focus of our radiocarbon dating project at Toca do Serrote da Bastiana, Brazil.*

Our research group at Texas A&M University has begun to directly radiocarbon date rock art pigments at several rock art sites in the Serra da Capivara National Park region in collaboration with Niéde Guidon. Using the plasma extraction method to date organic material in the paints, we now have radiocarbon dates on four other images in the Toca do Serrote da Bastiana shelter: 1880 ± 60, 2280 ± 110, 2970 ± 300, 3320 ± 50 (*29* and two previously unpublished dates). In addition, organic material extracted from the pigment of the red anthropomorph in question gave a radiocarbon age of 3730 ± 90 years BP.

## Background

In northeastern Brazil, the Serra da Capivara National Park region occupies an area of around 40,000 km$^2$ located in the southeast quadrant of the state of Piaui. On the edge of the Precambrian crystalline plain of the São Francisco River depression and the mountain ranges (*serras*) of the Permian-Devonian sedimentary basin of the Parnaiba River, the town of São Raimundo Nonato is set in an impressive landscape. For more than 200 km, cliffs formed by erosion with up to 250 m drops form a spectacular border between these two contrasting geological zones. Searches for archaeological sites have been limited to the area between S 8° to 9-1/2° latitude and W 41-1/5° to 43-1/2° longitude around the present day communities of São Raimundo Nonato, São João do Piaui, Canto do Buriti, Anisio de Abreu, and Caracol. The region's rich and diverse prehistoric art includes over 300 rock shelters containing paintings, engravings, or a combination of the two forms of expression. Certain shelters, like Toca do Boqueirão of Pedra Furada, harbor more than a thousand figures; and because many fragments of painted rock have been found buried in archaeological strata, it is estimated that 40% of the images have spalled away. However, in general, most sites are decorated with somewhere between ten and a hundred figures, but sometimes only one. With little doubt, rock art played an important role among prehistoric populations in Brazil.

With human occupation in Brazil possibly spanning as far back as 50,000 years ago, the 30,000-40,000 year old date of the Toca do Serrote da Bastiana rock painting measured by ESR and TL is feasible. Rock art images such as the 35,000-year-old paintings in Chauvet Cave in France demonstrate that rock art can remain visible for tens of thousands of years (*30*). However, it is important to note that European Paleolithic paintings are in protected underground cave environments, while the images in Brazil occur predominantly in open shelters more exposed to atmospheric weathering. In fact, the presence of painted spalled wall fragments in the excavated archaeological sediment suggests that the rock shelter surface is not a permanent canvas (*10*).

## Site description

The Toca do Serrote da Bastiana shelter is small, perhaps 3-4 meters wide, with an overhang of only a few meters. It is located at a limestone outcrop where local inhabitants were, until recently, actively dynamiting the rock to process it to lime by heating on-site. That activity has damaged – and had threatened to destroy completely – the rock paintings in the shelter. There are approximately a dozen images there.

## Sample Collection

During the summer of 2000, two of us (NG and MWR) visited the Toca do Serrote da Bastiana shelter for sampling. Visual examination of the ~20 cm tall red anthropomorphic motif dated by ESR and TL indicated that an area of about 1 $cm^2$ of the pictograph was coated with undisturbed 'calcite', the rest having been removed earlier for ESR and TL studies (*12*). The painting itself is a bright, apparently unfaded, red iron ochre color. Rubber gloves were worn throughout sampling and also during later handling in the laboratory to avoid modern contamination. A new surgical scalpel blade was used to remove the accretion layer, while we took care not to dig into the underlying paint pigment. It was easy to accomplish the removal of the accretion and it separated cleanly from the pictograph, leaving the pigment layer intact. The sample was scraped onto a clean square of aluminum foil, wrapped and placed into a sealable plastic bag. A sample of red paint from the same anthropomorphic figure as well paint samples from other images at Toca do Serrote da Bastiana were taken by Guidon and coworkers and then sent to Texas A&M for analysis.

## Experimental Methods

### Chemical Pretreatment

Procedures for chemically pretreating archaeological samples typically involve an acid-base-acid treatment (*31, 32*). However, we routinely eliminate both acid washes as we have shown them to be unnecessary with our plasma-chemical extraction technique. Carbonate and oxalate carbon are not extracted by the plasma; only organic material was removed for radiocarbon measurements (*33,34*). Samples are immersed in ~1 M NaOH and placed in an ultrasound bath for an hour at 50±5°C. When the resulting supernatant was colored, subsequent NaOH washes were performed until the supernatant appeared colorless because humic acids appear brownish-yellow in NaOH. Rinsed samples are then filtered through binder-free borosilicate glass filters baked overnight at ~600°C to remove organic contamination. Humic acids present will stay in solution and pass through the glass filter, removing unwanted organic contamination from the solid paint sample. Plasma extractions are run of dried filtrate material.

### Plasma Extraction of Carbon from Pigment Samples

The plasma-chemical extraction method has been used to separate organic carbon from rock art pigment samples for radiocarbon dating and has been described in detail elsewhere (*33-40*). The main advantage of this technique is the ability of the plasma to selectively oxidize organic carbon present in a sample and leave accompanying carbonates and oxalates intact (*33, 34*). Ultra-high purity bottled argon and oxygen (99.999 %) gases were used for all plasmas. A rotary pump and a turbo molecular pump maintain vacuum conditions (~$10^{-4}$ and ~$10^{-7}$ torr, respectively). First, low-temperature oxygen plasmas were used to pre-clean the reaction chamber; these were repeated until ≤0.001 mg carbon, as $CO_2$, was released. Samples were introduced into the chamber via a stainless-steel, copper-gasketed, flange port under a flow of argon to prevent atmospheric $CO_2$, aerosols, or organic particles from entering the system.

After the chamber was resealed and the sample partially degassed under vacuum, low temperature (<150°C) argon plasmas were used to desorb $CO_2$ molecules from the sample and chamber walls by inelastic collisions of the non-reactive, but energetic, argon species. The amount of adsorbed $CO_2$ and water on the sample was thus diminished to correspond to <0.001 mg carbon equivalent as is usual in our procedure.

Next, low-temperature (≤150°C), low-pressure (~1 torr) oxygen plasmas oxidized organic components of the sample to $CO_2$. Decomposition of inorganic carbon present (dolomitic limestone rock and calcite/calcium oxalate accretions) was prevented by running the plasmas at low-temperature. Carbon dioxide from the sample was flame-sealed into a glass tube cooled to liquid nitrogen temperature (-194°C), after water had been frozen out with a dry-ice/ethanol slurry (-58°C), and finally sent for radiocarbon analysis at the Center for Accelerator Mass Spectrometry at the Lawrence Livermore National Laboratory (LLNL-CAMS). It was necessary to utilize an AMS measurement due to the small sample size.

### Detection and Extraction of Calcium Oxalate

For the accretion covering the red anthropomorph, the presence of both calcium carbonate and calcium oxalate was confirmed by FTIR and powdered XRD. The principal minerals of the accretion layer were calcite (calcium carbonate) and whewellite (calcium oxalate monohydrate). Approximately 3 mg of the powdered sample were pressed into a KBr pellet for FTIR analysis on a Midac M2000 FTIR. The resulting spectrum shows absorption peaks consistent with the presence of oxalate, as well as a large carbonate component (Figure 2). The spectrum also indicates the presence of other constituents, probably silicates (clay or quartz). For XRD analysis, the sample was ground to a fine powder using an agate mortar and pestle. Calcite was determined to be the main constituent of the sample, with some whewellite present.

Due to the uncertainty of the relative amount of oxalate present, the entire remaining sample (~560 mg) was treated at Arkansas State University for the AMS $^{14}C$ and stable carbon isotope ratio ($\delta^{13}C$) analyses to maximize the probability that sufficient oxalate would be present for viable results. To remove carbonates, the powdered sample was placed in a Teflon beaker and treated with 25 ml of dilute phosphoric acid. After the initial acid/carbonate effervescence ceased, the pH of the solution was adjusted to pH ~ 2 using concentrated phosphoric acid. This pH was maintained for several days to ensure all carbonates were removed. This phosphoric acid treatment was tested and reported in Russ et al. (*25*). A final FTIR analysis established the removal of carbonate by the disappearance of the strong absorption peak at 1405 $cm^{-1}$ and also confirmed the presence of oxalate with absorption peaks at 1618 and 1315 $cm^{-1}$. The remaining sample was divided into two aliquots with approximately half sent to The University of Arizona AMS facility for $^{14}C$ analysis and the remaining half to T. W. Boutton at Texas A&M for the $\delta^{13}C$ measurement.

## Results

An aliquot of the processed sample contained 4.45% carbon, corresponding to 23.7 % $CaC_2O_4$, and only 0.02 % nitrogen. The small fraction of nitrogen implies a negligible amount of organic matter (except oxalate) in the accretion sample. The remainder of the sample is most likely silicates that do not affect a radiocarbon analysis. The $\delta^{13}C$ was measured to be -11.67 ‰, in the range of values determined for oxalate crusts in the Lower Pecos River region of southwest Texas (*25*). These results are compared to the $\delta^{13}C$ values for other materials in Table I. Hedges et al. (*41*) reported a similar $\delta^{13}C$ value of -10.3 ‰ for an oxalate crust over a rock painting in Argentina. The Brazilian oxalate $\delta^{13}C$ value suggests that neither marine carbonates ($\delta^{13}C$ ~0 ‰) nor unknown

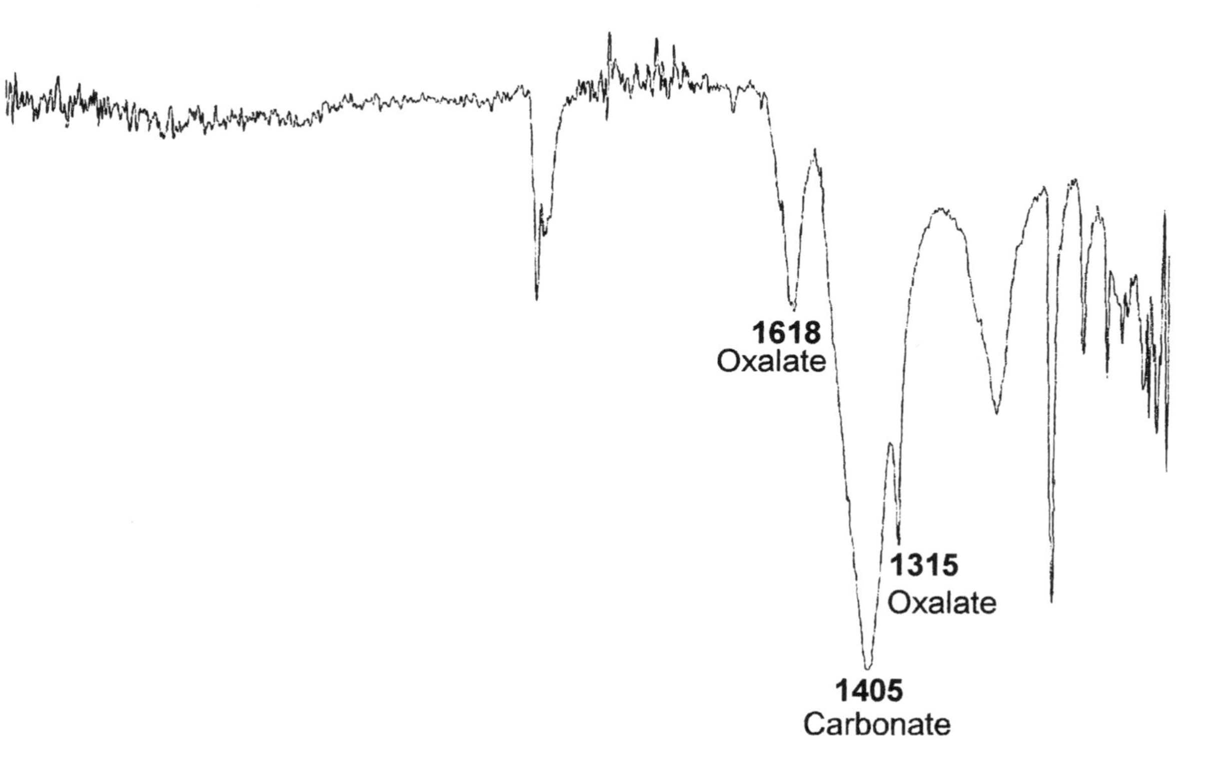


*Figure 2. FTIR spectrum showing presence of oxalate and carbonate in the accretion sample. After acid treatment, the strong absorption peak at 1405 $cm^{-1}$ disappeared indicating removal of carbonates.*

organic materials ($\delta^{13}C$ typically -25‰) will have a significant effect on dating. A radiocarbon measurement on the oxalate sample gave an age of 2,540 ± 60 years BP (AA-42663 on 07-07-01), corrected for a $\delta^{13}C$ of -11.7 ‰ measured at the NSF – University of Arizona AMS Facility. A second measurement on the same graphite target gave an age of 2,470 ± 40 years BP (AA-42664 on 08-05-01), corrected for a $\delta^{13}C$ of -11.6 ‰. These two radiocarbon measurements on the same sample yielded a combined age of 2490 ± 30 years BP.

**Table I. Approximate $\delta^{13}C$ values for various materials including the oxalate crust studied here**

| *Source* | *$\delta^{13}C$* | *Reference* |
|---|---|---|
| C3 plants | -27 ‰ | *(42)* |
| C4 plants | -13 ‰ | *(42)* |
| Woody tissue of C3 plants | -25 ‰ | *(32)* |
| Marine limestones | 0 ‰ | *(32)* |
| Atmospheric $CO_2$ (rural) | -7.8 ‰ | *(42)* |
| Oxalate crusts in Texas | -10.6 ‰[a] | *(25)* |
| Calcium oxalate sample (Boutton) | -11.67 ‰ | |
| Calcium oxalate sample (UA AMS) | -11.7 ‰ | |

[a] Average value of oxalate crusts from the Lower Pecos River region of southwest Texas.

The radiocarbon results for organic matter extracted from five different pictograph paint samples taken from Toca do Serrote da Bastiana, including the ~20 cm tall red anthropomorph, are shown in Table II. The red paints contain iron ochre and the black ones are charcoal. The high uncertainty associated with 4BR312 is due to small sample size (~0.020 mg of carbon). We would not normally report a radiocarbon age on such a small sample, but the measurement is consist with the other radiocarbon results and shows that the age of the pictograph is certainly not 30,000-40,000 years old. Typically, >0.050 mg of carbon is necessary to obtain a viable radiocarbon age.

**Table II. Radiocarbon Results of Paint Samples**

| *Texas A&M* | *AMS Laboratory* | *Description* | *Sample Size (mg C)* | *Radiocarbon Age (years BP)* |
|---|---|---|---|---|
| 4BR309 | CAMS-62934 | Red "batonnets" | .05 | 2280 ± 110 |
| 4BR310 | CAMS-62935 | Black sloth 1 | .12 | 1880 ± 60 |
| 4BR312 | CAMS-67685 | Black sloth 2 | .02 | 2970 ± 300 |
| 4BR320 | CAMS-68108 | Black figure | .22 | 3320 ± 50 |
| 4BR366 | CAMS-77890 | Red anthropomorph | .07 | 3730 ± 90 |

## Discussion

The radiocarbon date of 2490 years BP obtained on an oxalate accretion covering the red anthropomorph and the direct radiocarbon date on the same figure, as well as dates on four additional pictographs in the shelter, are all inconsistent with a date of 30,000-40,000 years ago determined by ESR and TL (*12*). Both dating methods, ESR/TL and oxalate radiocarbon dating, should yield equivalent dates corresponding to a similar event – the formation of a calcium carbonate and calcium oxalate accretion layer covering the pictograph. Both provide minimum ages for painting given that the pictograph must be at least as old as the calcite/oxalate crust covering it.

Re-crystallization and deposition processes most likely formed the calcite portion of the accretion when water containing soluble calcium carbonate coated the shelter wall. Formation of individual calcite layers would have occurred with multiple events throughout the entire history of the painting even up until the present time. ESR and TL measure accumulated electron traps produced as the newly formed calcium carbonate accretion is exposed to ionizing radiation. If the environmental radiation dose rate is known, then the time of crystal formation or when the accretion was deposited can be determined.

There are various proposed origins of calcium oxalate rock coatings that occur worldwide, including biological sources such as lichen, and bacteria, as well as chemical reactions of dissolved organic acids in rain aerosols at the rock/atmosphere boundary *(13-15, 17, 21-24)*. Nevertheless, the source of carbon is atmospheric, meaning that the carbon is contemporaneous with oxalate crust formation. Thus, the coating can be radiocarbon dated to determine when the oxalate was formed. Oxalate crust formation most likely occurred with multiple sequential events forming microscopic layers. A radiocarbon determination for an oxalate crust, therefore, would be a weighted average of the deposited layers' ages. With an extreme assumption of uniform oxalate deposition starting 3730 years ago and continuing to the present, the theoretical measured $^{14}C$ activity of 1790 BP, while not identical, is the same order of magnitude to our experimentally determined age of the oxalate accretion. In cases where such coatings are observed to cover pictograph paint layers, radiocarbon dating of the oxalate crust can constrain the production of the image by establishing a minimum age for the rock art (*22, 23, 25-28*).

Our oxalate radiocarbon determination is consistent with the era determined by radiocarbon dates on four other pictographs located in the same shelter. In fact, we have also obtained a direct radiocarbon date of 3730 ± 90 years BP on organic material extracted from the same red figure covered by the oxalate accretion. This plasma-chemical process has been verified by successfully dating known age materials and pictographs with archaeologically constrained ages (*39, 40)*.

How can the discrepancy between the two techniques, based on different assumptions, i.e., radiocarbon versus ESR/TL, be explained? Assuming for the moment that the ESR and TL dates are correct for the time of calcite crystallization, and thus the 'true' age of the oxalate should be in that general age range, we must conclude that both radiocarbon ages of the pigment and oxalate have been skewed drastically by young carbon. The most likely source of modern carbon would be biological microorganisms living in the crust. Only with the inclusion of ~73% modern carbon could we have obtained a measured date of 2490±30 years BP for the oxalate, assuming a 'true' age of 30,000-40,000 years. No bio-contamination was observed under optical microscopy and such a high amount of contamination is unlikely. We have not experienced that level of modern carbon contamination in our twelve years of radiocarbon dating.

If the oxalate radiocarbon age truly represents a minimum age of the painting, what can have affected the ESR and TL dates, making them so much older than ~2,500 years? Since ESR and TL are based on the same assumptions, agreement between the two is perhaps fortuitous. One key assumption in ESR and TL dating is that the dating signal is re-set to zero at the time of origin of the sample. In other words, the time of crystal formation corresponds to the time of interest to geologists and archaeologists (*43*). If as fresh calcite crystallized from solution and then deposited on the painting, some older particulate carbonate could have been incorporated into the liquid without dissolving. This inclusion of Precambrian calcite (>570 million years old) into freshly deposited calcite crystals would skew the ESR /TL dates to be older than the 'true' age. Since the crystallization age of the rock carbonate, possibly being incorporated as wind blown dust, would be millions of years old, a large effect could result. Some support for this scenario may be found in the presence of silicates that we saw in the FT-IR analysis, probably incorporated into the calcite/oxalate accretion as dust. In the Lower Pecos River region of Texas, Russ et al. (*25*) suggests that a silicate component in oxalate crusts is most likely the result of eolian dust adhering to surfaces when damp from dew or fog was present. The fraction of undissolved rock carbonate dust necessary for skewing the ESR and TL dates from 2,500 to 30,000-40,000 years would not be high because of the very ancient age of the rock. Another assumption of ESR and TL dating is that the environmental history of the sample needs to be determined (*43*). The uptake of radioisotopes such as uranium and thorium in the surrounding rock would increase the number of electron traps produced. Subsequent leaching of radioisotopes may have occurred. Even if the current concentration of radioisotopes in the rock substrate and accretion are accurately measured, it may not accurately reflect the past environment.

A reviewer was concerned with the possible inclusion of carbonate carbon during oxalate formation. We cannot totally rule out a fraction of the oxalate carbon arising from dissolved carbonate. But, limestone carbon cannot be the *sole* source of the oxalate carbon because of the C-14 content. The conversion of carbonate carbon to oxalate carbon would result in an oxalate radiocarbon age older than the 'true' age and our conclusion that the accretion layer is not 30,000-40,000 years old would still be valid. Russ et al. (*22, 23, 25*) and Watchman et al. *(26-28)* obtained radiocarbon results that suggest oxalate does not contain $^{14}$C-free carbon by the superposition of multiple oxalate and known age pictograph paint layers. Furthermore, the oxalate $\delta^{13}$C value of -11.67‰ is closer to atmospheric (-7.8‰) than to marine carbonates (~0 ‰).

## Conclusions

Unfortunately, our radiocarbon dates on the oxalate crust and organic matter in the paint layer cast doubt on the exciting ESR and TL dates obtained for the red anthropomorph pictograph at Toca do Serrote da Bastiana. An AMS radiocarbon date on an oxalate crust covering the red anthropomorph yielded a minimum date of 2490 ± 30 years BP for the painting. Another AMS radiocarbon age of 3730 ± 90 years BP on organic material in the underlying paint layer is consistent not only with the oxalate result, but with four direct radiocarbon dates obtained on organic matter extracted from other rock paintings in the same shelter.

## Acknowledgments

KLS was partially supported by a Regent's Fellowship from the office of the Vice-Provost for Research at Texas A&M University. Additional funding came from the Fundação Museu do Homen Americano, Brazil, and the Program to Enhance Scholarly and Creative Activities, Office of the Vice-President for Research at Texas A&M University.

## References


1. Guidon, N.; Delibrias, G. *Nature* **1986**, *321*, 769-771.
2. Guidon, N.; Arnaud, B. *World Archaeology* **1991**, *23*, 167-178.
3. Meltzer, D. J.; Adovasio, J. M.; Dillehay, T. D. *Antiquity* **1994**, *68*, 695-714.
4. Guidon, N.; Pessis, A. M.; Perenti, F.; Fontugue, M.; Guérin, C. *Antiquity* **1996**, *70*, 408-421.
5. Haynes, C. V. *Science* **1969**, *166*, 709-715.
6. Bahn, P. G. *Nature* **1993**, *362*, 114-115.
7. Borrero, L. A. *Antiquity* **1995**, *69*, 602-603.
8. Dennell R.; Hurcombe, L. *Antiquity* **1995**, *69*, 604.
9. Prous, A.; Fogaça, E. *Quaternary International* **1999**, *53/54*, 21-41.
10. Guidon, N. *Natural History* **1987**, *96*, 6-12.
11. Daltrini Felice, G. *Sitio Toca do Boqueirao da Pedra Furada, Piaui –Brasil: Estudo comparativo das estratigrafias extra sitio*, Univdersidad Federal de Pernambuco: Recife, Brasil, 2000.
12. Watanabe, S. *Personal communication,* 2001.
13. Del Monte, M.; Sabbioni, C. *Studies in Conservation* **1987**, *32*, 114-121.
14. Del Monte, M.; Sabbioni, C.; Zappia, G. *The Science of the Total Environment* **1987**, *67*, 17-39.
15. Watchman, A. L. *Rock Art Research* **1990**, *7*, 44-50.
16. Edwards, H. G. M.; Farwell, D. W.; Seaward, M. R. D. *Spectrochemica Acta* **1991**, *474*, 1531-1539.
17. Watchman, A. L. *Studies in Conservation* **1991**, *36*, 24-32.
18. Scott, D. H.; Hyder, W. D. *Studies in Conservation* **1993**, *38*, 155-173.
19. Watchman, A. L.; Sale, K.; Hogue, K. *Conservation and Management of Archaeological Sites* **1995**, *1*, 25-34.
20. Hyman, M.; Rowe, M. W. *Science Spectra* **1998**, *13*, 22-27.
21. Lazzarini, L; Salvadori, O. *Studies in Conservation* **1989**, *34*, 20-26.
22. Russ, J.; Palma, R. L.; Loyd, D. H.; Boutton, T. W.; Coy, M. A. *Quaternary Research* **1996**, *46*, 27-36.
23. Russ, J.; Kaluarchi, W. D.; Drummond, L.; Edwards, H. G. M. *Studies in Conservation* **1999**, *44*, 91-103.
24. Bonaventura, M. P. D.; Gallo, M. D.; Cacchino, P.; Ercole, C.; Lepidi, C. *Geomicrobiology* **1999**, *16*, 55-64.
25. Russ, J.; Loyd, D. H.; Boutton, T. W. *Quaternary International* **2000**, *67*, 29-36.
26. Watchman, A. *Geoarchaeology* **1993**, *8*, 465-473.
27. Watchman, A.; Cambell, J. *2nd International Symposium on the oxalate films in the conservation of works of art, Milan* **1996**, 409-422.
28. Watchman, A.; David, B.; McLiven, I. J.; Flood, J. M. *Journal of Archaeological Science* **2000**, *27*, 315-325.

29. Waltman, E.; Steelman, K.; Rowe, M. W. *Sociedade Brasileira De Pesquisadores Nikkeis Scientific Journal* **2000**, *4*, 90-97.
30. Clottes, J. In *Experiment and Design: Archaeological Studies in Honour of John Coles*; Harding, A. F., Ed.; Oxbow Books: Oxford, UK, 1999, pp 13-19.
31. Taylor, R. E. *Radiocarbon Dating: An Archaeological Perspective*; Academic Press: New York, NY, 1987; pp 120-123.
32. Bowman, S. *Radiocarbon Dating*; Interpreting the Past, University of California Press/British Museum: Berkeley, CA, 1990, pp 20-23.
33. Russ, J.; Hyman, M.; Rowe, M. W. *Radiocarbon* **1992**, *34*, 867-72.
34. Chaffee, S. D.; Hyman, M.; Rowe, M. W. In *New Light on Old Art: Recent Advances in Hunter-Gather Rock Art*; Whitley, D. S.; Loendorf, L. L., Eds.; Monograph 36; Institute of Archaeology, University of California: Los Angeles, CA, 1994, pp 9-12.
35. Russ, J.; Hyman, M.; Shafer, H. J.; Rowe, M. W. *Nature* **1990**, *348*, 710-711.
36. Hyman, M.; Rowe, M. W. In *American Indian Rock Art*; Freers, S. M., Ed.; American Rock Art Research Association: San Miguel, CA, 1997; Vol. 23, pp 1-9.
37. Rowe, M. W. In *Handbook of Rock Art Research*; Whitley, D. S., Ed.; Altamira Press: New York, NY, 2001; pp 139-166.
38. Pace, M. F. N.; Hyman, M.; Rowe, M. W.; Southon, J. R. In *American Indian Rock Art*; Bock, F. G., Ed.; American Rock Art Research Association: Tuscon, AZ, 2000; Vol. 24, pp 95-102.
39. Armitage, R. A.; Brady, J. E.; Cobb, A.; Southon, J. R.; Rowe, M. W. *American Antiquity* **2001**, 66, 471-480.
40. Hyman, M.; Rowe, M. W. *Techne* **1997**, *5*, 61-70.
41. Hedges, R. E. M.; Ramsey, C. B.; Van Klinken, G. J.; Pettitt, P. B.; Nielsen-Marsh, C.; Etchegoyen, A.; Fernandez Niello, J. O.; Boschin, M. T.; Llamazares, A. M. *Radiocarbon* **1998**, *40*, 35-44.
42. Boutton, T. W. In *Carbon Isotope Techniques*; Coleman, D. C.; Fry, B., Eds.; Academic Press: San Diego, CA, 1991, pp 173-185.
43. Skinner, A. R. *Applied Radiation and Isotopes* **2000**, *52*, 1131-1136.

# Chapter 4

# Measuring Lead Isotope Ratios in Majolica from New Spain Using a Nondestructive Technique

**Susan Reslewic and James H. Burton**

**Laboratory for Archaeological Chemistry, Department of Anthropology, University of Wisconsin, Madison, WI 53706**

We report a new nondestructive technique using EDTA extraction and Inductively Coupled Plasma Mass Spectrometry (ICP-MS) to measure lead isotope ratios among majolica sherds from six 18th century presidios in northern New Spain. Our preliminary results support previous archaeological and chemical studies that suggest one or two production centers as the source of majolica found in New Spain. This research has broad relevance for research on lead-glazed wares, and raises specific questions regarding the production and distribution of majolica in 18th century New Spain.

Archaeological work at 17th – 18th century presidios along the northern frontier of New Spain has yielded large amounts of lead-glazed pottery known as majolica (*1,2*). Majolica pottery consists of an earthenware clay body, surrounded by a tin-opacified glaze to which lead was added as a flux (*3*-5). The origin of the majolica technology lies in the Old World (*3,4,6*), where the majolica became popular in Italy and Spain by the 13th century. Accompanying the arrival of the Spanish to the New World in the early 16th century was the establishment of majolica manufacturing centers, first in Mexico City, and later, in Puebla (*6-9*). In addition to New World majolica manufacture, majolica from Spain continued to be exported to New Spain until the 18th century (*10*).

Majolica typologies based on paste attributes and painted surface decoration have been used to construct majolica sequences throughout New Spain as well as to differentiate between majolica made in Mexico City and in Puebla (*2,5,11*). Archaeometric studies on majolica have complemented typological approaches, and have been crucial in determining where majolica found in New Spain was manufactured. For example, Olin et al. (*12*) report that concentrations of the elements lanthanum, cerium, and thorium in the clay paste can distinguish majolica manufactured in Spain from Spanish styles of majolica manufactured in the New World. In a later study, Olin et al. (*9*) determined that majolica sherds from Mexico City and Puebla could be distinguished based on their chromium, iron, and scandium concentrations. A study by Magetti et al. (*13*) further showed in addition to elemental concentrations, majolica from Spain, Mexico City, and Puebla could be resolved according to mineralogical signatures in the pottery's temper. With respect to the glaze, lead isotopes have been shown to be useful in determining the provenience of majolica sherds. Using Thermal Ionization Mass Spectrometry (TIMS), Joel et al. (*14*) determined that $^{208}Pb/^{206}Pb$ and $^{207}Pb/^{206}Pb$ ratios can distinguish between Spanish and Mexican majolica.

This study inspired our research on the utility of lead isotopes to learn about the the provenience of majolica found at the presidios of Northern New Spain. From the late 16th to late 18th centuries, the Spanish set up fortified settlements, called presidios, in an effort to protect the northern border of New Spain (*1*). Over time, the presidios lost their strictly military flavor and became the heart of a civilian center, home to Spanish soldiers, merchants, farmers, and their families. Spaniards living at the presidios exerted a high demand for majolica (*1,2,6,8*). Yet, wagon train deliveries of supplies from settled areas of Mexico were unreliable, in part due to the difficult and dangerous route through unoccupied territory to the northern borderlands (*5*). It would seem economically profitable for majolica manufacturing industries to be established at or near the frontier sites in order to achieve a constant supply of majolica without the problems of long distance transport. In our attempt to study the provenience of majolica from frontier sites based on lead isotope analyses, we also hoped to explore the possibility that people living at the presidios had set up one or a few of their own manufacturing industries in order to circumvent the logistical difficulties of acquiring majolica from a major manufacturing center such as Puebla.

Our research goals led us to develop a nondestructive technique for measuring lead isotopes in majolica, having potential for various kinds of lead-glazed pottery. After soaking majolica sherds in a dilute solution of ethylenediaminetetraacetic acid (EDTA), we used Inductively Coupled Plasma Mass Spectrometry (ICP-MS) to measure the resulting lead isotope ratios in the EDTA-Pb chelate. Our results are consistent with previous archaeological and chemical studies that have cited one or two Mexican majolica manufacturing centers as the key supply source for majolica throughout New Spain in the 18$^{th}$ century. More important, as our technique is aesthetically nondestructive, the method described here has the potential to be used on a variety of glazed pottery types, particularly from museum contexts in which nondestructive analyses are preferred and often required.

## Methods and Results

The majolica samples used in this study were donated from the Arizona State Museum, courtesy of Dr. Michael Jacobs. The sherds are from surface collections at six frontier sites in what is now Arizona and Sonora (Figure 1). Table 1 lists the six sites, their locations, extent of occupation, and the number of sherds from each site.

**Table I. Context of Majolica Samples**

| *Site* | *Location* | *Site Dates* | *Number of Sherds* |
|---|---|---|---|
| Tuscon | Pima Co., Arizona | ? – AD 1890 | 7 |
| Terrenate | Cananea Mun., Sonora | AD 1742 – AD 1775 | 29 |
| Fronteras | Fronteras Mun., Sonora | AD 1690 - ? | 7 |
| San Bernardino | Agua Prieta Mun., Sonora | AD 1775 – AD 1790 | 7 |
| Tubac | Santa Cruz Co., Arizona | AD 1730 – AD 1865 | 3 |
| Guevavi | Santa Cruz Co., Arizona | AD 1691 – AD 1775 | 12 |

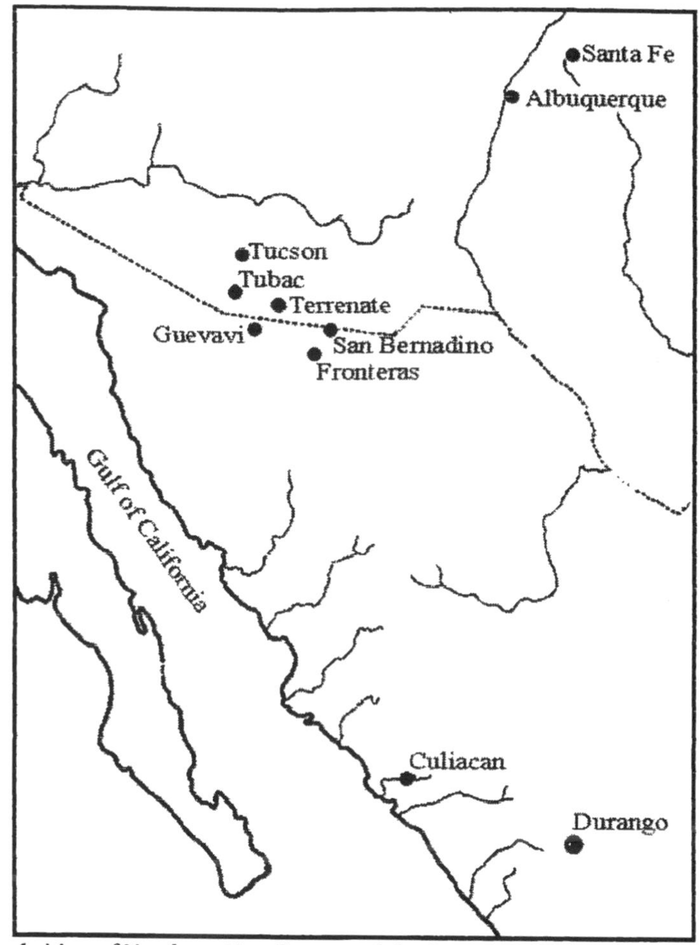


*Figure 1. Map of Northern New Spain and the Six Presidios (Modified from reference 1. Copyright 1975 University of Oklahoma Press)*

We soaked each sherd in a 0.025% EDTA solution for fourteen days, which was sufficient time to extract lead concentrations in the range of $10^1$-$10^2$ parts per billion (ppb). In addition, it is likely that the sherds could soak for less than fourteen days, either by using a saturated solution of EDTA (0.05%), or using a warm EDTA solution which would allow a greater EDTA solubility. We chose EDTA due to its well known ability to chelate with a variety of positively charged metal ions (*15*). With four strongly acidic carboxyl protons and two less acidic ammonium protons, EDTA acts as a multidentate ligand that forms strong 1:1 complexes with most metal ions. Soaking majolica sherds in a solution of

EDTA causes the extraction of $Pb^{2+}$ ions from the glaze and the formation of an EDTA-Pb complex (Figure 2) with no apparent destruction to the pottery's glaze or clay body.

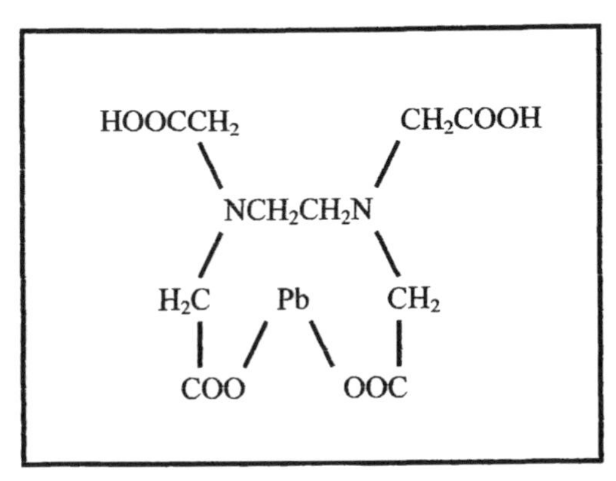


*Figure 2. EDTA-Pb Chelate*

In addition to analyzing the EDTA-Pb solution with the ICP-MS, as a control we also analyzed several solutions of the EDTA solution after sitting for fourteen days in containers without pot sherds. The near absence of lead in our control samples gave us confidence that the lead detected by the ICP-MS came from the majolica glaze rather than from the plasticware in which the sherds were soaked.

Table II shows the precision and accuracy of the ICP-MS for the four lead isotopes $^{204}Pb$, $^{206}Pb$, $^{207}Pb$, and $^{208}Pb$ based on measurements of the NIST Common Lead Isotopic Standard (SRM 981). For all of the lead isotopes except for $^{204}Pb$, our precision is under 0.2%, and our percent error, or accuracy, is roughly 0.2%. This is expected, as $^{204}Pb$ is the least abundant lead isotope, and in our analyses we measured amounts of $^{204}Pb$ that were smaller than counts for the other isotopes by approximately a factor of ten. In addition, previous research used $^{207}Pb/^{206}Pb$ versus $^{208}Pb/^{206}Pb$ to identify clusters of majolica pottery manufactured in different areas (*14*), and it is with these isotopes that our method has its best precision and accuracy.

The lead isotope ratios we measured for the majolica samples are shown in Table III. Figure 3 shows a graph of $^{207}Pb/^{206}Pb$ versus $^{208}Pb/^{206}Pb$ for the samples from every site. The range of lead isotope ratios fall roughly within the range generated by Joel et al (*14*) for majolica produced in both Spain and

**Table II. Precision and Accuracy**

| | $^{204}Pb/^{208}Pb$ | $^{206}Pb/^{208}Pb$ | $^{207}Pb/^{208}Pb$ | $^{206}Pb/^{207}Pb$ |
|---|---|---|---|---|
| NIST SRM 981 (actual) | 0.0272 | 0.4612 | 0.4220 | 1.0930 |
| NIST SRM 981 (experimental) | | | | |
| #1 | 0.0277 | 0.4614 | 0.4214 | 1.0950 |
| #2 | 0.0276 | 0.4630 | 0.4224 | 1.0963 |
| #3 | 0.0276 | 0.4618 | 0.4224 | 1.0934 |
| #4 | 0.0274 | 0.4612 | 0.4207 | 1.0964 |
| Mean | 0.0276 | 0.4619 | 0.4217 | 1.0953 |
| Standard Deviation | 0.0001 | 0.0008 | 0.0008 | 0.0014 |
| % RSD (Precision) | 0.4125 | 0.1747 | 0.1964 | 0.1272 |
| % Accuracy | 1.359582 | 0.146339 | -0.071875 | 0.208467 |

Mexico. However, there is no good evidence for any clustering, inter- or intra-site, based on lead isotope ratios of the majolica found from the six presidios.

Lead isotope ratios for the twenty nine sherds from Terrenate span almost the entire range of lead isotope ratios for the majolica from all sites. The nine Terrenate sherds at the upper right corner of Figure 3 *may* represent a distinct group of majolica, as they have the highest values of both $^{208}Pb/^{206}Pb$ and $^{207}Pb/^{206}Pb$. Figure 4 shows the spread of lead isotope ratios for majolica sherds only from Terrenate. One could argue that this graph shows a resolution of the Terrenate sherds into two groups based on their lead isotope ratios. However, in the absence of lead isotope data from geological sources in the area, it is difficult to attribute these potential "groups" to different proveniences, without knowing the geological differences among lead isotope sources in the region. A greater sample size and information on lead isotope ratio values among potential geological sources of lead will be necessary to determine whether or not people at Terrenate acquired majolica raw materials or pottery from more than one place.

**Table III. Lead Isotope Ratios of Majolica Samples**

| *Sample* | $^{204}Pb/^{208}Pb$ | $^{206}Pb/^{208}Pb$ | $^{207}Pb/^{208}Pb$ | $^{206}Pb/^{207}Pb$ |
|---|---|---|---|---|
| Tubac | | | | |
| SR3 | 0.0264 | 0.4844 | 0.4049 | 1.1962 |
| SR6 | 0.0262 | 0.4853 | 0.4069 | 1.1925 |
| SR7 | 0.0264 | 0.4855 | 0.4063 | 1.1950 |
| Fronteras | | | | |
| SR8 | 0.0258 | 0.4794 | 0.4029 | 1.190 |
| SR9 | 0.0261 | 0.4835 | 0.4048 | 1.194 |
| SR10 | 0.0261 | 0.4808 | 0.4040 | 1.190 |
| SR11 | 0.0262 | 0.4825 | 0.4049 | 1.192 |
| SR12 | 0.0258 | 0.4850 | 0.4013 | 1.209 |
| SR13 | 0.0262 | 0.4826 | 0.4035 | 1.196 |
| SR14 | 0.0261 | 0.4826 | 0.4047 | 1.192 |
| San Bernardino | | | | |
| SR15 | 0.0259 | 0.4797 | 0.4029 | 1.191 |
| SR16 | 0.0258 | 0.4831 | 0.4025 | 1.200 |
| SR17 | 0.0263 | 0.4835 | 0.4051 | 1.194 |
| SR18 | 0.0260 | 0.4826 | 0.4038 | 1.195 |
| SR19 | 0.0260 | 0.4819 | 0.4033 | 1.195 |
| SR20 | 0.0262 | 0.4861 | 0.4058 | 1.198 |
| SR21 | 0.0261 | 0.4831 | 0.4043 | 1.195 |
| Guevavi | | | | |
| SR22 | 0.0262 | 0.4838 | 0.4053 | 1.194 |
| SR23 | 0.0261 | 0.4841 | 0.4036 | 1.199 |
| SR24 | 0.0261 | 0.4824 | 0.4032 | 1.197 |
| SR25 | 0.0260 | 0.4825 | 0.4040 | 1.194 |
| SR26 | 0.0279 | 0.4717 | 0.4162 | 1.133 |
| SR27 | 0.0260 | 0.4825 | 0.4032 | 1.197 |
| SR28 | 0.0262 | 0.4832 | 0.4048 | 1.194 |
| SR29 | 0.0258 | 0.4812 | 0.4025 | 1.195 |
| SR30 | 0.0262 | 0.4827 | 0.4038 | 1.196 |
| SR31 | 0.0263 | 0.4859 | 0.4051 | 1.200 |
| SR32 | 0.0263 | 0.4858 | 0.4052 | 1.199 |
| SR33 | 0.0262 | 0.4821 | 0.4048 | 1.191 |
| SR34 | 0.0260 | 0.4816 | 0.4040 | 1.192 |

**Table III. Lead Isotope Ratios of Majolica Samples (cont'd)**

| | | | | |
|---|---|---|---|---|
| Terrenate | | | | |
| SR 35 | 0.02544 | 0.47904 | 0.40298 | 1.18874 |
| SR36 | 0.02531 | 0.47725 | 0.40027 | 1.19233 |
| SR37 | 0.02570 | 0.47803 | 0.40189 | 1.18945 |
| SR38 | 0.02538 | 0.47749 | 0.39978 | 1.19437 |
| SR39 | 0.02583 | 0.48148 | 0.40286 | 1.19515 |
| SR40 | 0.02564 | 0.47708 | 0.40368 | 1.18183 |
| SR41 | 0.02546 | 0.47591 | 0.40203 | 1.18376 |
| SR42 | 0.02567 | 0.47878 | 0.40408 | 1.18485 |
| SR43 | 0.02561 | 0.47838 | 0.40339 | 1.18589 |
| SR44 | 0.02513 | 0.47525 | 0.40007 | 1.18792 |
| SR45 | 0.02613 | 0.48463 | 0.40453 | 1.19801 |
| SR46 | 0.02623 | 0.48484 | 0.40516 | 1.19665 |
| SR47 | 0.02638 | 0.48580 | 0.40606 | 1.19636 |
| SR48 | 0.02623 | 0.48367 | 0.40496 | 1.19436 |
| SR49 | 0.02619 | 0.48484 | 0.40616 | 1.19372 |
| SR50 | 0.02619 | 0.48385 | 0.40389 | 1.19798 |
| SR51 | 0.02602 | 0.48361 | 0.40372 | 1.19787 |
| SR52 | 0.02579 | 0.48156 | 0.40268 | 1.19589 |
| SR53 | 0.02610 | 0.48163 | 0.40419 | 1.19160 |
| SR54 | 0.02623 | 0.48422 | 0.40432 | 1.19762 |
| SR55 | 0.02621 | 0.48416 | 0.40376 | 1.19912 |
| SR56 | 0.02627 | 0.48461 | 0.40557 | 1.19489 |
| SR57 | 0.02626 | 0.48279 | 0.40433 | 1.19405 |
| SR58 | 0.02633 | 0.48610 | 0.40533 | 1.19925 |
| SR59 | 0.02640 | 0.48547 | 0.40562 | 1.19688 |
| SR60 | 0.02639 | 0.48538 | 0.40599 | 1.19556 |
| SR61 | 0.02648 | 0.48648 | 0.40567 | 1.19919 |
| SR62 | 0.02628 | 0.48394 | 0.40471 | 1.19576 |
| SR63 | 0.02610 | 0.48455 | 0.40465 | 1.19745 |
| Tuscon | | | | |
| SR64 | 0.02641 | 0.48789 | 0.40687 | 1.19914 |
| SR65 | 0.02611 | 0.48399 | 0.40460 | 1.19621 |
| SR66 | 0.02615 | 0.48509 | 0.40515 | 1.19731 |
| SR67 | 0.02626 | 0.48534 | 0.40545 | 1.19702 |
| SR68 | 0.02624 | 0.48406 | 0.40473 | 1.19600 |
| SR69 | 0.02627 | 0.48498 | 0.40345 | 1.20210 |
| SR70 | 0.02647 | 0.48680 | 0.40544 | 1.20069 |

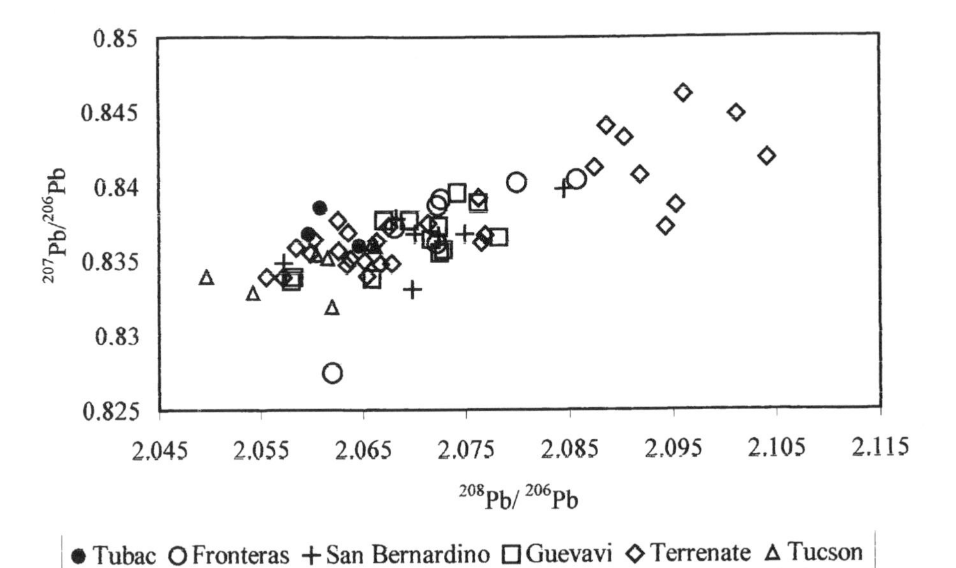


*Figure 3. Graph of* $^{207}Pb/^{206}Pb$ *versus* $^{208}Pb/^{206}Pb$ *for majolica samples.*

## Discussion

Our method for measuring lead isotope ratios in glazed pottery using EDTA extraction followed by ICP-MS has allowed us to study majolica from the northern frontier of New Spain with no apparent destruction to the artifacts. Our results are congruent with previous research on lead isotope ratios in majolica pottery and our values for $^{207}Pb/^{206}Pb$ and $^{208}Pb/^{206}Pb$ fall within the range reported for both Spanish and Mexican majolica (*14*). In addition, our results support previous studies that have cited a few production centers as responsible for supplying the people of New Spain with majolica.

There are several possible reasons for our inability to identify distinct majolica groups, based on lead isotope ratios, among the pottery from the presidio sites. Perhaps, despite the logistical difficulties of transporting supplies to northern New Spain from settled areas of Mexico to the south, people living at the presidios relied on the major production centers, particularly at Puebla, for their majolica supply. Although there was a high demand for majolica at the presidios (*2*), majolica was a prestigious item and a status symbol with which the Spanish closely identified (*6*). There may have

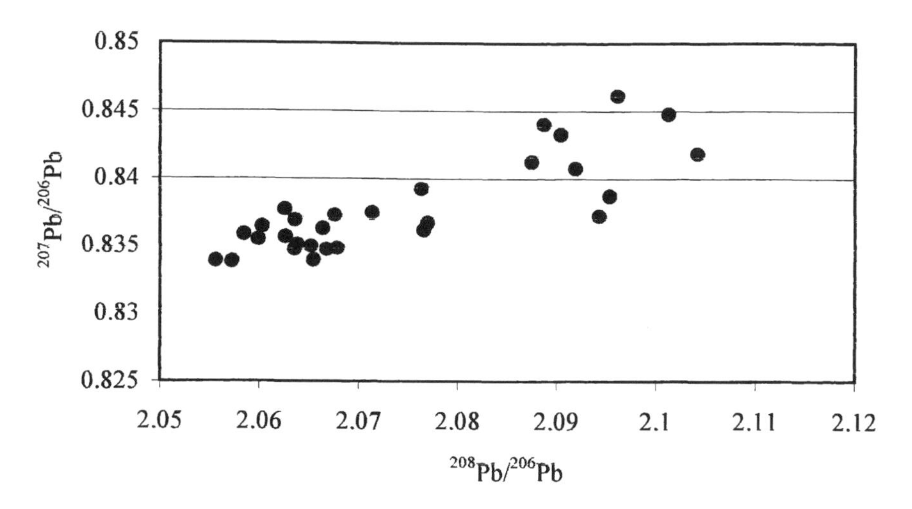


*Figure 4. Graph of $^{207}Pb/^{206}Pb$ versus $^{208}Pb/^{206}Pb$ for Terrenate samples.*

been political or practical reasons preventing the establishment of several small scale production centers that could efficiently and reliably supply the northern frontier sites with majolica. In addition, we know from research by Lister and Lister (*6*) and Goggin (*5*) that the majolica industry at Puebla was highly organized and rigidly controlled. There may have been explicit or implicit barriers for majolica making industries elsewhere in New Spain, particularly along the frontier.

However, it is also possible that lead isotope studies provide insufficient resolution for identifying chemical differences in majolica from the presidios. Further research on lead isotope ratio differences among geological deposits in the region are key for determining whether or not lead isotopes are appropriate in an attempt to study differences in majolica production among sites in northern New Spain. Finally, a larger sample size would be helpful to understand the variation in majolica lead isotope ratios among the presidios, and would aid in the interpretation of the Terrenate data. Carlson et al (*16*) used NAA analysis of 33 elements in Texas Mission ceramics to differentiate between ceramics manufactured in Mexico and Texas, and determined that people at Texas missions were making glazed pottery. Future archaeological and chemical research will be key in learning whether or not people living at or near the presidios were manufacturing majolica and relating this information to our image of life at the frontier of New Spain.

## Conclusions

EDTA extraction followed by ICP-MS is a nondestructive technique for measuring lead isotopes in majolica, and has potential to be used on other varieties of lead-glazed pottery. Our results from studying majolica sherds from sites at the northern frontier of New Spain are congruent with previous studies on majolica, in terms of both lead isotope ratio numerical values and suggestions that a few majolica manufacturing centers supplied New Spain with majolica. We hope that our nondestructive method will make a greater proportion of artifacts, particularly from museums, available for chemical studies, and will inspire future research on the role of chelating agents like EDTA in archaeometry research.

## Acknowledgments

We wish to thank Dr. Michael Jacobs of the Arizona State Museum for loaning us the majolica samples used in this study. We thank Dr. T. Douglas Price of the University of Wisconsin-Madison for supporting this project. In addition, we wish to thank Dr. Pamela Vandiver of the Smithsonian Institution's Center for Materials Research and Education for her insights on this project, her help with the preliminary literature research, and for getting us in touch with scholars who have done chemical studies of majolica.

This research was supported by the Laboratory for Archaeological Chemistry in the Department of Anthropology at the University of Wisconsin-Madison.

## References

1. Moorhead, M. L. *The Presidio: Bastion of the Spanish Borderlands*; University of Oklahoma Press: Norman, OK, 1975.
2. Cohen-Williams, A. G. *Historical Archaeology* **1987**, *26*, 119-130.
3. Lambert, J. B. *Traces of the Past: Unraveling the Secrets of Archaeology Through Chemistry*; Perseus Books: Reading, MA, 1997.
4. Rice, P. M. *Pottery Analysis: A Sourcebook*; University of Chicago Press: Chicago, IL, 1987.
5. Goggin, J. S. *Spanish Majolica in the New World: Types of the Sixteenth to Eighteenth Centuries;* Yale Univ. Publications in in Anthropology No. 72, Yale University Press: New Haven, CT, 1968.

6. Lister, F. C.; Lister R. H. *Andalusian Ceramics in Spain and New Spain: A Cultural Register from the Third Century BC to 1700*; University of Arizona Press: Tucson, AZ, 1987
7. Lister, F. C.; Lister, R. H. *Historical Archaeology* **1984**, *18*, 87 – 102.
8. Lister, F. C.; Lister, R. H. *Sixteenth Century Majolica Pottery in the Valley of Mexico;* Anthropological Papers of the University of Arizona No. 39: University of Arizona Press: Tucson, AZ, 1982.
9. Olin, J. S.; Blackman, M. J. In *Archaeological Chemistry IV*; Allen, R. O., Ed.; Advances in Chemistry Series No. 220; American Chemical Society: Washington, DC, 1989; pp 87 – 112.
10. Myers, J. E.; de Amores Carredano, F.; Olin, J. S.; Hernandez, A. P. *Historical Archaeology* **1987**, *26*: 131 – 147.
11. Deagan, K. *Artifacts of the Spanish Colonies of Florida and the Caribbean, 1500 – 1800, Volume I: Ceramics, Glassware, and Beads*; Smithsonian Institution Press: Washington, DC, 1987.
12. Olin, J. S.; Harbottle, G.; Sayre, E. V. In *Archaeological Chemistry II*; Carter, G. F., Ed.; Advances in Chemistry Series No. 171; American Chemical Society: Washington, DC, 1978; pp. 200 – 229.
13. Magetti, M.; Westley, H; Olin, J. S. In *Archaeological Chemistry III;* Lambert, J. B., Ed.; Advances in Chemistry Series No. 205; American Chemical Society: Washington, DC, 1984; pp. 151 – 191.
14. Joel, E. C.; Olin, J. S.; Blackman, M. J.; I. L. Barnes. In *Proceedings of the 26th International Archaeometry Symposium*. Farquhar R. M.; Hancock, R. G. V.; Pavlish, L. A.; Eds; University of Toronto Archaeometry Laboratory: Toronto, Canada, 1988, pp. 188 – 185.
15. Harris, D. C. *Quantitative Chemical Analysis*, 4th ed.; W. H. Freeman and Company, NY, 1995.
16. Carlson, S. B.; James, W. D., Carlson, D. L. *Compositional Analysis of Spanish Colonial Ceramics: The Texas Missions*. Paper Presented at the 1994 Annual Meeting of the Society for Historical Archaeology, Vancouver, British Columbia, Canada, 1994.

# Chapter 5

# Characterization of Archaeological Materials by Laser Ablation–Inductively Coupled Plasma–Mass Spectrometry

**Robert J. Speakman, Hector Neff, Michael D. Glascock, and Barry J. Higgins**

**Research Reactor Center, University of Missouri, Columbia, MO 65211**

Use of inductively coupled plasma-mass spectrometry (ICP-MS) coupled with a laser ablation sample introduction system (LA-ICP-MS), as a minimally destructive method for characterization of archaeological materials, has steadily increased during the past five years. Although method development still is in its early stages, LA-ICP-MS has demonstrated to be a productive avenue of research for chemical characterization of obsidian, chert, pottery, and painted and glazed surfaces. LA-ICP-MS applications and comparisons with other analytic techniques are described.

A variety of analytic techniques currently are used to provide chemical characterizations of archaeological materials. These techniques which include instrumental neutron activation analysis (INAA), X-ray fluorescence (XRF), proton induced X-ray emission (PIXE), X-ray diffraction (XRD), scanning electron microscopy (SEM), inductively coupled plasma-atomic emission spectroscopy (ICP-AES), inductively coupled plasma-mass spectrometry (ICP-

 

MS) have made significant contributions to archaeology in terms of understanding details of technology, organization of production, and interaction patterns. While each of these methods has its advantages, ICP-MS which has become available during the last decade has emerged as a versatile and powerful analytic tool for compositional-based studies of archaeological materials (*1-6*).

Our purpose in this paper is to provide an overview of ICP-MS and present case studies demonstrating how LA-ICP-MS can be used to characterize obsidian, chert, ceramic glazes, and pigments. It is our hope that the research we discuss herein will inspire future archaeological projects in which data obtained using LA-ICP-MS assists archaeologists in understanding past cultural phenomenon.

## Introduction to LA-ICP-MS

ICP-MS is a relatively new technique for determining the chemical compositions for a wide variety of liquid and solid sample matrices with high precision, accuracy, and sensitivity. Since the introduction of the first commercial instruments in 1983 (*7-8*), ICP-MS has evolved into one of the most powerful analytic tools available for determining elemental concentrations at ultra trace levels. The introduction of laser ablation as a sample introduction method in 1985 (*9*) effectively expanded the capabilities of ICP-MS so that not only could bulk chemical characterizations be obtained through digestion of a sample in acid but also specific loci of a sample could be targeted and characterized using a laser.

Over the past five years archaeologists have begun to use ICP-MS with increasing regularity to address archaeological questions (*10-13*). At the Missouri University Research Reactor (MURR), we have used LA-ICP-MS to characterize a variety of archaeological materials. These projects include characterization of paints on pottery from the Mesa Verde region of the American Southwest (*14*),Turkey and Peru; analysis of glazes on Mesoamerican Plumbate pottery (*15*), historic Spanish and Mexican Majolica pottery (*16*), and historic Euroamerican pottery; characterization of glass beads, Midwestern cherts (*17*), obsidian (*18-19*), turquoise and determination of inclusions in pottery. These studies clearly demonstrate the potential of LA-ICP-MS for addressing archaeological questions.

Unlike INAA, XRF, or ICP-MS of solutions which produces a bulk elemental characterization of the entire matrix, LA-ICP-MS provides a point specific characterization of the ablated area of the sample. On one hand, attempts to obtain compositional data for heterogeneous samples such as pottery can be challenging because of inherent spatial variation in the sample (e.g., clay and temper particles). On the other hand, reduction of the laser beam diameter, enables inclusions in heterogeneous areas to be characterized. Relatively homogeneous samples, such as obsidian and to a certain extent cherts, paints, and glazes, are ideally suited for LA-ICP-MS since spatial variation is minimal in these materials. ICP-MS can provide compositional data for 50-60 elements including lanthanides, whereas, other techniques typically provide compositional data for about 30 (or less) different elements. Some elements such as lead which cannot be measured by INAA but can be measured by LA-ICP-MS may prove important for separating materials into different compositional groups. For many elements LA-ICP-MS has lower detection limits than other instrumental techniques (e.g., Sr, Sb, Ba, Zr).

In LA-ICP-MS, the sample is placed inside a sample holder or laser cell where ablation takes place. The ablated area varies in size depending on the sample matrix but is usually smaller than 1000 X 1000 μm and less than 30 μm deep. The ablated material is flushed from the laser cell using a 1.1-1.3 liter/minute flow of argon or an argon/helium-mixed carrier gas through Tygon tubing, and introduced into the ICP-MS torch where an argon gas plasma capable of sustaining electron temperatures between 8000 and 10,000 K is used to ionize the injected sample. The resulting ions are then passed though a two stage interface designed to enable the transition of the ions from atmospheric pressure to the vacuum chamber of the ICP-MS system. Once inside the mass spectrometer, the ions are accelerated by high voltage and pass through a series of focusing lenses, an electrostatic analyzer, and finally a magnet. By varying the strength of the magnet, ions are separated according to mass/charge ratio and passed through a slit into the detector which records only a very small atomic mass range at a given time. By varying the instrument settings the entire mass range can be scanned within a short period of time.

The instrument used in the studies reported here is a Thermo Elemental Axiom magnetic-sector inductively coupled plasma mass spectrometer capable of resolving atomic masses as close as 0.001 atomic mass units apart, thus eliminating many interferences caused by molecular ions that pose problems for quadrupole ICP-MS instruments. The ICP-MS is coupled to a Merchantek Nd:YAG 213-nanometer wavelength laser ablation unit. The laser can be targeted on spots as small as five micrometers in diameter. The small spot size and the high sensitivity of magnetic-sector ICP-MS to a wide range of major, minor, and trace elements make LA-ICP-MS a very powerful microprobe. Moreover laser ablation is virtually non-destructive to most samples considering that the ablated areas are often indistinguishable with the naked eye. LA-ICP-

MS is also comparable in safety to SEM and XRF instruments, provided that the factory-installed safety features are not bypassed by the instrument operator.

### Analytic Parameters

Each day, prior to data acquisition, the instrument is turned-on and permitted to warm-up for a minimum of one hour. Warming-up the instrument permits internal components to reach their optimum operating temperature, greatly reducing instrument noise and drift. After an hour, a glass standard (SRM-612) is placed in the laser cell and continuously ablated to produce a signal that permits the instrument settings to be tuned so that signal intensity and stability are maximized. After tuning the instrument, data for a blank and a combination of standards are collected, usually NIST SRM 610 and 612 (glass wafers doped with 61 elements), Glass Buttes obsidian, and Ohio Red clay tiles fired to 1200°C. While Glass Buttes obsidian is not a certified reference material it has been characterized by a variety of analytic techniques, including LA-ICP-MS, making it ideal for use as a standard (*10*). Ohio Red clay has been characterized at MURR by INAA in excess of 300 times also making it suitable for use as a LA-ICP-MS standard. Following collection of the standard data, the unknown samples are placed in the laser cell and analyzed. Additional sets of standards are collected periodically throughout the day.

## Characterization of Obsidian

Our experiments at MURR have shown that obsidian is perhaps the easiest material to characterize using LA-ICP-MS and that these data are comparable to elemental data determined by other analytic methods (*10,20*). In an early experiment, data for three Jemez Mountain (New Mexico) obsidian source samples previously analyzed by INAA and XRF (*21*) were reanalyzed using LA-ICP-MS and ED-XRF (energy dispersive XRF) (*19*). A raster approximately 400 $\mu m^2$ was placed over a relatively flat area of each sample. The laser was operated at 80% power using a 200 µm diameter beam, firing 20 times per second. The laser was set to scan across the raster at a speed of 30 µm per second. The laser beam was permitted to pass over the ablation area one time prior to data acquisition in order to 1) remove possible contamination from the surface of the sample, and 2) permit time for sample uptake and for the argon gas plasma to stabilize after the introduction of the fresh ablation material. Our experiments have demonstrated that pre-rastering permits the laser to better couple with the sample matrix. Analytes of interest were scanned three times and averaged. In most cases, the %RSD was less than 10%.

Monitoring the amount of material removed by the laser and transported to the ICP is complicated making normalization of data difficult. Conditions such as the texture of the sample, location of the sample in the laser cell, surface topography, laser energy, and other factors affect the amount of material that is introduced to the ICP torch and thus the intensity of the signal monitored for the various atomic masses of interest. In addition, instrumental drift affects count rates. With liquid samples internal standards typically are used to counteract instrument drift, but this approach is not feasible when material for the analysis is ablated from an intact solid sample. If one or more elements can be determined by another analytic technique, then these can serve as internal standards. In the case of rhyolitic obsidian, which has relatively consistent silicon concentrations (ca. 36%), we have determined that silicon count rates can be normalized to a common value. Likewise, standards are normalized to their known silicon concentrations. This value, divided by the actual number of counts produces a normalization factor from which all the other elements in that sample can be multiplied. A regression of blank-subtracted normalized counts to known elemental concentrations in the standards yields a calibration equation that can be used to calculate elemental concentrations in the samples analyzed.

When normalized to silicon, the LA-ICP-MS data obtained for the Jemez Mountain obsidian closely correlate to INAA and ED-XRF data (Figure 1). Having developed a method for normalization and standardization that apparently worked for North American obsidian source samples, we applied this method to a collection of artifacts from Honduras submitted for INAA analysis. The ablation parameters for this experiment were identical to those used for the Jemez experiment with the exception that the laser power settings were reduced to 60% instead of the original 80%. Data obtained from this experiment also were normalized to silicon and converted to abundance data as described previously. In this case, the analyst (Speakman) had no prior knowledge of what source groups these samples had been assigned based on the INAA data. The LA-ICP-MS data were evaluated using quantitative approaches routinely employed at MURR (*22*). When the LA-ICP-MS compositional groups were compared to the INAA compositional groups, we realized a 100% success for classification of the LA-ICP-MS samples. INAA source data were then plotted in the same log base-10 elemental and principal component space as the LA-ICP-MS data to obtain a measure of accuracy for this method (Figure 2-3). As expected, there was a close agreement between the INAA and LA-ICP-MS data. The main differences observed between these two analytic techniques result from higher INAA detection limits for barium. In this case, the INAA Pachuca source data does not correlate with the LA-ICP-MS Pachuca artifact data as closely as

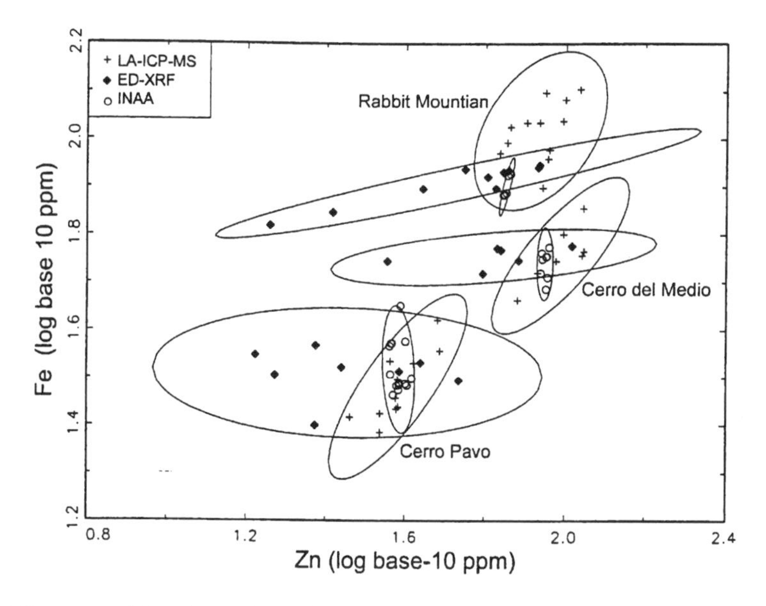


Figure 1. Bivariate plot of zinc and iron log-concentrations for Jemez Mountain, NM obsidian source groups characterized by LA-ICP-MS, INAA, and ED-XRF. Ellipses represent 90% confidence levels for group membership.

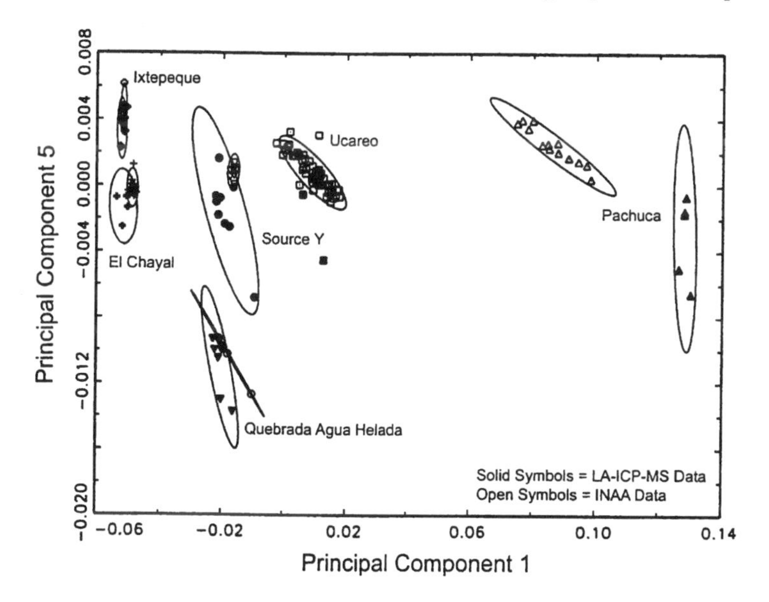


Figure 2. Plot of principal components 1 and 2 for Honduran obsidian artifacts analyzed by LA-ICP-MS and Mesoamerican obsidian source groups characterized by INAA. LA-ICP-MS data are normalized to silicon. Ellipses represent 90% confidence levels for group membership.

we would like as a result of the different detection limits for each analytic technique (Figure 2).

One problem with normalizing obsidian data to silicon is that if the instrument settings vary significantly from day to day, particularly if the instrument is routinely changed from solid sample introduction to liquid sample introduction, mass-bias effects can be introduced into the data as a result of different operational settings for the instrument. The consequence of this is that data for the same compositional groups collected on different days can exhibit high standard deviations. One of us, (Neff) developed a method for normalization of LA-ICP-MS data in matrices that could not be normalized to an internal standard (e.g., glazes, pottery, and pigments). Neff determined that reasonably accurate results could be obtained if the total elemental counts were normalized from each sample to a single standard value (e.g., 10 million counts for all standards and unknowns). A regression of blank-subtracted normalized counts to known elemental concentrations in the standards then yields a calibration equation that can be used to calculate elemental concentrations in the unknowns[1].

The total counts normalization method was applied to the same Honduran obsidian data set (Figure 4). The results, while yielding distinct compositional groups agreed poorly with the INAA source data. When the instrument settings and ablation parameters were changed in our next obsidian characterization project to more closely reflect those used by Neff (*15*), we obtained results that were comparable to INAA data (*18*). In this particular experiment, obsidian artifacts submitted from an archaeological site in Guatemala were analyzed for Ba, Cs, Hf, Th, and Zr by LA-ICP-MS. The resolving power was set to 4000 for all elements even though some of the elements have little or no polyatomic interferents. Laser power was increased to 80% and we used a 400 μm diameter beam instead of a 200 μm diameter beam. The laser repetition rate remained at 20 times per second, and the scan speed was increased from 30 to 70 μm per second. Increasing the laser power for this experiment resulted in increased instrument sensitivity. Increasing the size of the laser diameter permits ablation of additional material which serves to average out variation that may occur in the sample matrix. Our experiments further indicated that by increasing the scan speed from 30 to 70 μm per second we could increase sensitivity and signal stability.

In addition to the analysis of the 180 samples submitted for this project, seven samples from known obsidian sources were analyzed to provide additional verification of our source identifications since these samples would not be analyzed by other instrumental methods. Elemental counts per second obtained

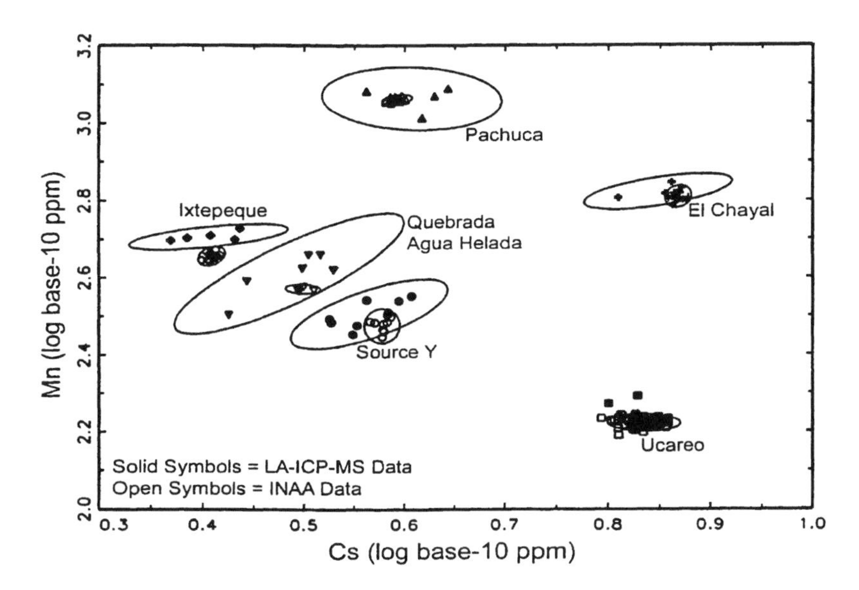


Figure 3. Bivariate plot of cesium and manganese log-concentrations for Honduran obsidian artifacts analyzed by LA-ICP-MS and Mesoamerican obsidian source groups characterized INAA. LA-ICP-MS data are normalized to silicon. Ellipses represent 90% confidence levels for group membership.

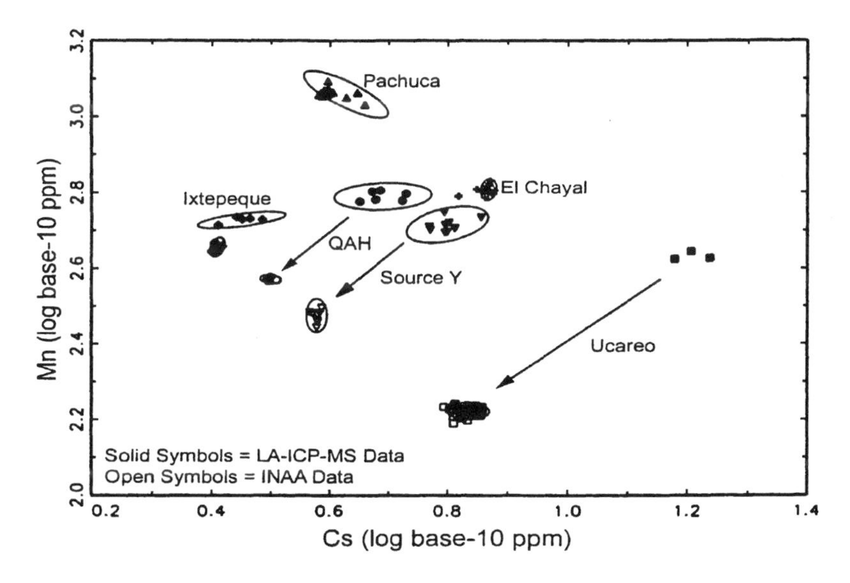


Figure 4. Bivariate plot of cesium and manganese log-concentrations for Honduran obsidian artifacts analyzed by LA-ICP-MS and Mesoamerican obsidian source groups characterized INAA. LA-ICP-MS data are normalized to total counts. Ellipses represent 90% confidence levels for group membership.

for the Guatemalan obsidian were normalized to total counts. A regression of normalized counts to known elemental concentrations in the standards provided a calibration equation that was used to calculate elemental concentrations in the unknowns. The artifacts from this experiment clustered into three distinct groups. Their source identification was verified by the LA-ICP-MS analysis of the seven known source samples. INAA source data plotted against the LA-ICP-MS data further demonstrated the precision and accuracy of this normalization method (Figure 5).

In summary, we have determined LA-ICP-MS analysis to be an accurate and precise instrumental technique for chemical characterization of obsidian artifacts and source samples. We have developed two methods for normalization and standardization of obsidian data that while different, permit LA-ICP-MS data to be transformed to abundance data and that data derived by LA-ICP-MS compares favorably to INAA data. The minimal sample preparation time, the ability to select the suite of elements that will be analyzed, and the rapid analytic time make LA-ICP-MS, a minimally invasive, cost-effective method for characterizing obsidian.

## Characterization of Chert

Chemical characterization of cherts has met with varying degrees of success throughout the years. INAA has traditionally been the preferred analytic technique as a result of low INAA detection limits for several elements. However, high compositional variation both within and between sources make chert difficult to characterize regardless of analytic technique. Additionally, low concentrations for many elements further complicate the analysis since potentially important elements for discriminating between compositional groups may occur in concentrations below INAA detection limits. In this example, four Missouri chert sources previously analyzed by INAA were reanalyzed for approximately fifty elements using LA-ICP-MS. A raster approximately 750 $\mu m^2$ was placed over a relatively flat area of the sample. The laser was operated at 80% power using a 200 μm diameter beam, firing 20 times per second. The laser was set to scan across the raster area at a speed of 30 μm per second. Each sample was pre-ablated once prior to data acquisition to remove possible surface contamination. Data were normalized to blank-subtracted silicon counts, and a regression of normalized counts to known elemental concentrations in the standards provided a calibration equation that was used to calculate elemental concentrations in the unknowns.

The INAA data demonstrated that each source group was compositionally distinct (Figure 6). Using LA-ICP-MS we hoped to ascertain whether the sensitivity of LA-ICP-MS was sufficient for characterizing cherts. Additionally, we were interested in determining if the source groups would remain compositionally distinct and how these data would compare to INAA data. We determined that characterization of chert using LA-ICP-MS was just as difficult as characterization by INAA due to similar reasons: the heterogeneous nature of the sample matrix and low elemental concentrations. However, in spite of these factors, the chert source groups remained distinct (Figure 7) indicating that LA-ICP-MS is an effective analytic technique for characterization of cherts.

Since characterization of obsidian and cherts using LA-ICP-MS has been demonstrated to be a productive avenue of research, it is just as probable that analysis of other lithic materials would prove to be advantageous. For example, pipestone artifacts which hold ceremonial and religious significance for certain Native American groups could be analyzed without significantly impacting the integrity of the artifact. Steatite, jasper, marble, and basalt are also possibilities where LA-ICP-MS analysis may provide important details for interpreting past cultural activity.

## Analysis of Paints and Glazes

One of the potentially more productive avenues of research using LA-ICP-MS involves the characterization of paints and glazes used in decoration of pottery. Just as bulk analysis of clays by INAA, XRF, ICP-MS, and other methods has demonstrated to be a productive avenue of research for making interpretations regarding past cultural systems, chemical characterization of paints and glazes may prove just as important. Historically, compositional analyses of paints and glazes has been restricted due to limitations imposed by the available analytic techniques. Wet chemistry, physical tests, SEM, PIXE, XRF and other instrumental techniques have been used to characterize paints and glazes with varying degrees of success. However, with INAA, for example, the painted or glazed surface must be separated from the ceramic matrix for analysis. Additionally, analysis of paint by INAA requires a sample size that would result in the destruction of large quantities of paint or glaze. With instrumentation that uses X-rays, it is difficult ensure the underlying clay matrix is not contributing to the analysis. Using LA-ICP-MS, painted and glazed surfaces can be analyzed without having to remove the painted or glazed decoration from the ceramic sherd. By adjusting laser settings the underlying clay matrix can be avoided thus permitting the painted and glazed surfaces to be characterized. Surface contamination can also be removed by using the laser to ablate the decorated

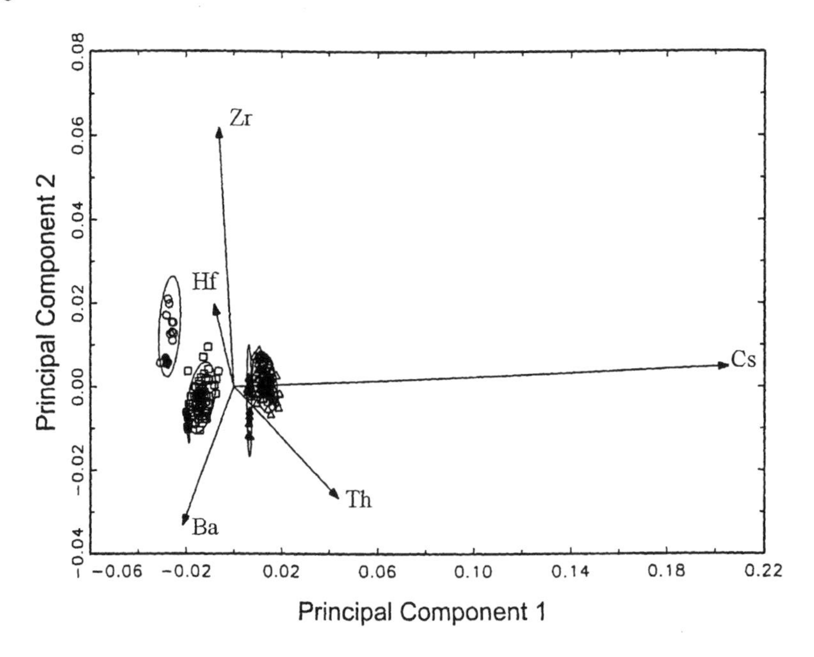


Figure 5. Plot of principal components 1 and 2 for Guatemalan obsidian artifacts analyzed by LA-ICP-MS and INAA obsidian source groups. LA-ICP-MS data are normalized to total counts. Ellipses represent 90% confidence levels for group membership.

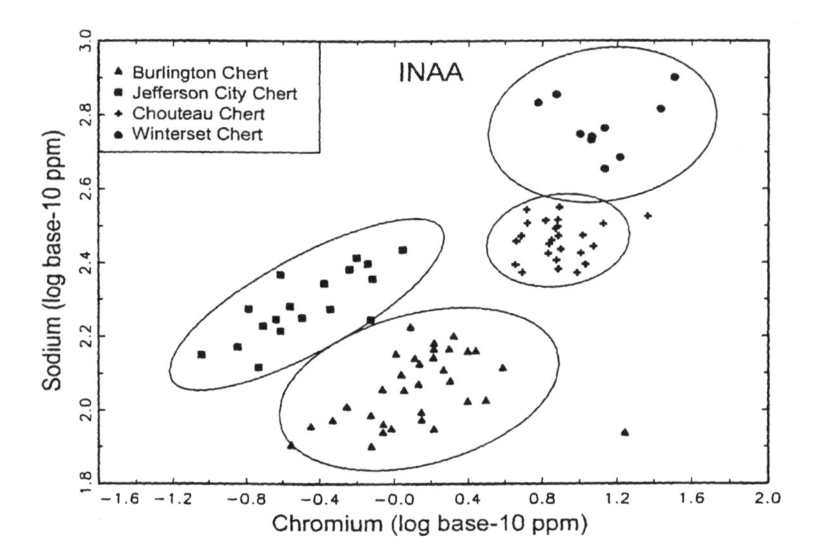


Figure 6. Bivariate plot of chromium and sodium log-concentrations for chert source groups analyzed by LA-ICP-MS. LA-ICP-MS data are normalized to silicon. Ellipses represent 90% confidence levels for group membership.

surface prior to data acquisition. Additionally, minimal sample preparation time, high sample throughput, and high instrument sensitivity and accuracy make LA-ICP-MS ideal for characterizing this sample matrix.

In a recent project, we analyzed prehistoric Puebloan pottery from the Mesa Verde Region of the American Southwest to determine the effectiveness of LA-ICP-MS for discriminating between the two major ceramic types, Mancos and Mesa Verde Black-on-white (*14*). Classification of these ceramics is based primarily on their assumed paint composition. Mancos Black-on-white is typically classified as having a mineral-based paint derived from an iron-manganese ore. The other ceramic type, Mesa Verde Black-on-white is usually described as having an organic-based paint. Temporally, Mancos Black-on-white precedes Mesa Verde Black-on-white, thus these two ceramic types are used as temporal markers and serve to test a variety of hypotheses surrounding technology, production, distribution, and social organization. Accurate classification of these pottery types provides a foundation for testing these hypotheses.

We analyzed a sample of both Mancos and Mesa Verde Black-on-white and obtained data from both the painted and unpainted white slipped areas of the pottery. A raster area approximately 900 $\mu m^2$ was placed over a relatively flat area on each sample. The laser was operated at 50% power to prevent the laser from burning through the paint and into the underlying clay matrix. We used a 100 μm diameter beam, firing 20 times per second and a laser scan rate of 30 μm per second. Lacking a suitable method for normalizing and standardizing data obtained from this experiment, we determined these data could best be expressed as a ratio of counts obtained from the painted area to counts obtained from the unpainted area of the same sherd[2]. The resulting ratio was then used to assign the paint into mineral or carbon-based paint groups (Figure 8). In our experiment, samples containing high iron and manganese ratios were classified as being mineral-based whereas paints with low iron and manganese ratios were assumed to be carbon-based. In addition to collecting data for iron and manganese, we also collected data for several other elements including copper, zinc, lead, and silver. High ratios of these elements in several samples indicate potentially greater variation exists in paint than previously thought.

In a separate experiment, we used LA-ICP-MS to characterize glazes on historic period ceramics recovered from nineteenth century Mormon pottery production workshops in Utah. This project, while limited in scope, demonstrated that clear glazes in Mormon-made pottery are highly variable. Some pottery assumed to be salt or lead glazed was determined to have high tin and antimony concentrations. This project served as an important reminder about

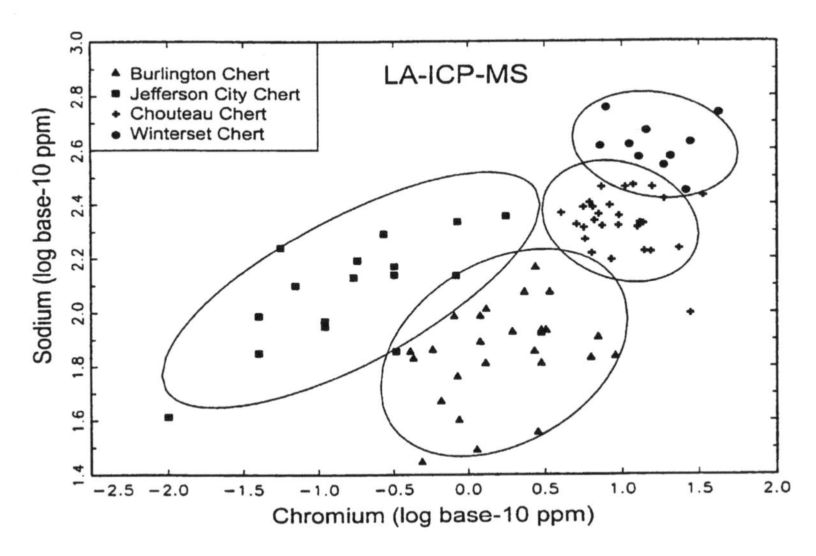


Figure 7. Bivariate plot of chromium and sodium log-concentrations for chert source groups analyzed by INAA. Ellipses represent 90% confidence levels for group membership.

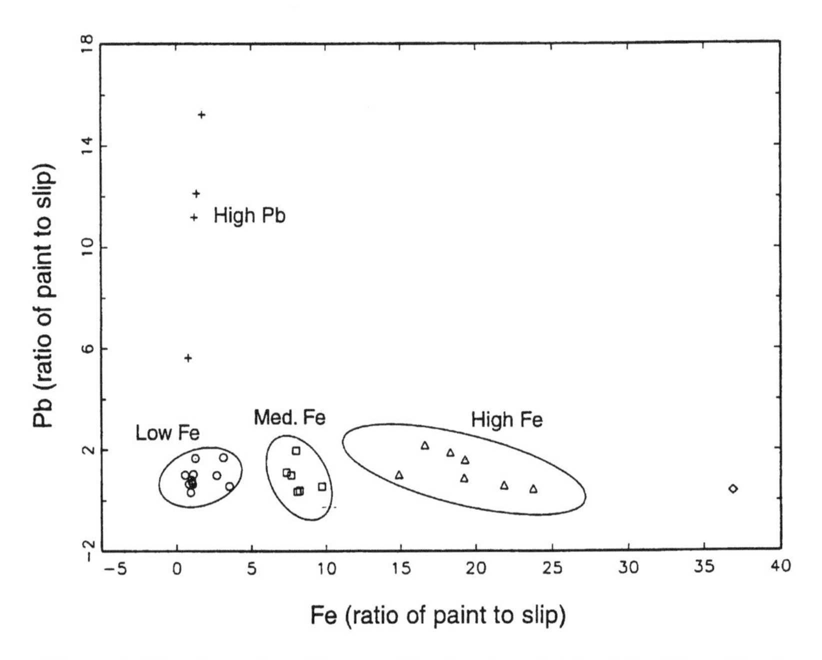


Figure 8. Bivariate plot of iron and lead ratios obtained for Mesa Verde and Mancos Black-on-white pottery. Ratios are based on counts obtained from the black painted area of a sherd to counts obtained for the slipped portion of the same sherd. Ellipses represent 90% confidence levels for group membership.

making determinations regarding composition based on visual attributes. For archaeologists interested in studying historic period ceramic technology, analysis of glazes may prove important for understanding ceramic modes of production and distribution.

LA-ICP-MS analysis of Colonial Period Spanish and Mexican glazed pottery from Mexico City similarly revealed interesting results. For this study we analyzed a large sample of tin and lead glazed pottery. The results of this study demonstrate that three tin-based compositional groups and two lead-based compositional groups were present in the sample (*16*). Additionally, we determined lead isotopic ratios for the lead glazed pottery. While preliminary, the isotopic data suggest that LA-ICP-MS may effectively discriminate between lead ores. It is possible that precision for isotopic data could be increased in future studies by using quadrupole or time of flight (TOF) ICP-MS. This project demonstrates that LA-ICP-MS can readily discriminate between different lead and tin-based compositional groups, indicating that classification of pottery into catch-all tin and lead categories may not be sufficient for fully exploring variability that may be present in archaeological assemblages. Elemental characterization of lead and/or tin glazes can reveal potentially important details of production and interaction.

## Conclusions

In this paper we have demonstrated how LA-ICP-MS can be used to characterize archaeological materials. While our discussion has centered upon characterization of lithics and painted and glazed surfaces, LA-ICP-MS has many other applications. For example, LA-ICP-MS has been used to characterize archaeological metals (*13, 23*), lake sediments (*24*), and human remains (*25*). As we continue to develop and refine our analytic procedures we anticipate that analyses of other archaeological materials will also be possible. Outside of archaeology, researchers continue to develop applications for LA-ICP-MS that may eventually see use by archaeologists. For example, several researchers have analyzed fish otoliths by LA-ICP-MS (*26-27*). Since the elemental and isotopic composition of fish otoliths reflect that of the environment, it is not hard to imagine that archaeologists will one day use compositional data derived from fish otoliths in conjunction with pollen and phytolith data to reconstruct prehistoric environments. LA-ICP-MS, while relatively new to archaeology and other scientific fields, will undoubtedly become one of the major analytic techniques used to characterize archaeological materials during the next 10-20 years. The range of materials that may be characterized by LA-ICP-MS is only limited by the imagination. What we have touched upon in this paper is only the beginning of a whole new field of archaeological research.

## Acknowledgements

The research reported here was funded in part by a grant from the National Science Foundation (SBR-9977237). The authors are indebted to Candace Sall for able assistance with sample preparation and analysis. We also thank Donna Glowacki, Jim Judge, Jack Ray, Enrique Rodriguez, Ed Schortman, and Natasha Tabares, for submitting some of the samples analyzed by LA-ICP-MS and discussed in this paper. Kathryn Jakes, LuElla Parks and an anonymous reviewer provided valuable comments on earlier drafts of this paper.

### Notes

1. While the total counts method for normalization and standardization of data yields results that are comparable to INAA, we since have adopted the method advocated by Gratuze and his colleagues (*10*) for normalization and standardization of LA-ICP-MS data (*15*).

2. At the time this paper was written, ratios seemed to work best for presenting paint and glaze data obtained by LA-ICP-MS. We have since adopted the method advocated by Gratuze and his colleagues (*10*) for normalization and standardization of LA-ICP-MS data and determined that this method works for these sample matrices (*15*) thus permitting paint and glaze data to be expressed as abundance data rather than ratios.

## References

1. Angelini, E.; Atzeni, C.; Bianco, P.; Rosalbino, F; Virdis, P. F., In *Applications of Plasma Source Mass Spectrometry II*; Holland, G.; Eaton, A. N.; Eds.; Royal Society of Chemistry, Cambridge, 1993; pp 165-174.
2. Mallory-Greenough, L.M.; Greenough G.D.; Owen, J.V. *J. Arch. Sci.* **1998**, 25, 85-87.
3. Pingitore, N. E.; Leach, J.; Villalobos, J.; Peterson, J.; Hill, D., In *Material Issues in Art and Archaeology V;* Vandiver, P.B.; Druzik, J.R.; Merkel, J.F.; Stewart, J.; Eds.; *Materials Research Society Symposium Proceedings,* 1997; pp 59-70.

4. Tykot, R. H. In *Archaeological Obsidian Studies*, M.S. Shackley Ed.; Plenum Press: New York, 1998; pp 67-82.
5. Tykot, R. H.; Young, S. M., In *Archaeological Chemistry: Organic, Inorganic, and Biochemical Analysis;* Orna, M. V., Ed.; ACS Symposium Series 625; Amer. Chem. Soc: Washington, DC, 1996; pp 116-130.
6. Young, S. M.; Phillips, D. A.; Mathien, F. J.; In Archaeometry 94: *The Proceedings of the 29th International Symposium on Archaeometry, Ankara 9-14 May, 1994;* Demirci, S.; Ozer, A. M.; Summers, G. D.; Eds.; pp 147-150.
7. Longerich, P. L., In *Laser Ablation-ICPMS in the Earth Sciences;* P. Sylvester, Ed.; Volume 29, Mineralogical Association of Canada, St. Johns, 2001; pp 21-28.
8. Montaser, A. Ed.; *Inductively Coupled Plasma Mass Spectrometry,* Wiley-VCH, New York, 1998.
9. Gray, A. *Analyst* **1985**, 110, 551-556.
10. Gratuze, B.; Blet-Lemarquard M.; Barrandon J. N. *J. Rad. Analy. Nuc. Chem* **2001**, 247, 645-656.
11. Gratuze, B. *J. Arch. Sci.* **1999**, 13, 869-881.
12. Kennett, D. J.; Neff, H.; Glascock, M. D.; Mason, A. Z.; *SAA Arch Record* **2001**, 1, 22-26
13 Devos, W.; Senn-Luder, M.; Moor, C.; Salter, C.; *Fresnius' J. Anal. Chem.* **2000**, 366, 873-880.
14. Speakman, R. J.; Neff, H. *Amer. Antiq*. **2002**, 67, 136-144.
15. Neff, H. *J. Arch. Sci.* **2002,** in press.
16. Rodriguez-Alegria, E. Ph.D. Thesis, University of Chicago, Chicago, **2002**.
17. Speakman, R. J.; Glascock, M. D.; Ray, J. Poster Presented at the Annual Meeting of the Society for American Archaeology, New Orleans, 2001
18. Speakman, R.J.; Neff, H.; Glascock, M.D.; Laser Ablation ICP-MS Analysis of Obsidian Artifacts from the Late Formative Ujuxte Site, Pacific Coast, Guatemala, Research Report, University of Missouri Research Reactor, 2001.
19. Sall, C.A.; Glascock, M.D.; Speakman, R.J. Poster Presented at the Annual Meeting of the Society for American Archaeology, New Orleans, 2001
20. Glascock, M.D. *Int. Assoc. Obsid. Stud.* **1999**, 23, 13-25.
21. Glascock, M.D.; Neff, H.; Stryker, K.S.; Johnson, T.N. *J. Rad. Nuc. Chem.* **1994,** 180, 29-35.
22. Neff, H., In *Introduction to Archaeological Sciences*, Brothwell, D.R.; Pollard, A.M.; Eds.; Wiley and Sons: New York, 2001, pp 733-747.
23. Junk, S. A. *Nuc. Inst. Meth. Phy. Res. B.* **2001**, 181, 723-727.
24. Eastwood, W.J.; Pearce, J.G. *J. Arch. Sci.* **1998**, 25, 677-687.
25. Budd, P.; Montgomery, A.C.; Krause, P.; Barriero, R.G. *Sci. Total Env.* **1998**, 222,121-136.
26. Campana, S.E. *Mar. Ecol. Prog. Ser.* **1999**, 188,263-297
27. Thorrold, S.R.; Jones, C.M.; Campana, S.E. *Limnol. Oceanogr.* **1997**, 42:102-111.

# Chapter 6

# Modern and Ancient Resins from Africa and the Americas

**Joseph B. Lambert[1,*], Yuyang Wu[1], and Jorge A. Santiago-Blay[2]**

**[1]Department of Chemistry, Northwestern University, 2145 Sheridan Road, Evanston, IL 60208-3113**
**[2]Department of Paleobiology, National Museum of Natural History, Smithsonian Institution, MRC 121, 10th Street and Constitution Avenue, NW, Washington, DC 20560**

New samples of resinous materials from Africa and the Americas, both modern and fossilized, have been characterized by solid state carbon-13 nuclear magnetic resonance spectroscopy. The large *Hymenaea*-related grouping has been expanded to include sites in North Africa (Morocco), West Africa (Mauritania), and several countries in South America (Guyana, Brazil, and Uruguay). The worldwide *Agathis*-related grouping has been slightly expanded to include locations in North Dakota, North Carolina, and Spain. The other worldwide group, possibly associated with Dipterocarpaceae, has been found to occur also in North Carolina, the second American site to be represented. We have found that *Dacryodes excelsa* (Burseraceae) from Puerto Rico exhibits the same spectra as materials from the Burseraceae family from Mexico and that *Guiacum officinale* (Zygophyllaceae) exhibits a unique new spectral pattern.

Resinous materials have played an important role in many past civilizations. They constitute one of the most important classes of organic artifacts. Fossilized resins, known in the European context as amber, are attractive, easily worked materials. Humans have been creating a variety of artifacts ranging from beads to elaborate statues and have been trading these materials since the Mesolithic period (approximately 5-10 thousand years ago). Modern resins also have served as adhesives, as preservatives (added for example to stabilize or flavor wine), as medicinals, and as ritual materials such as incense (*1*).

The scientific examination of amber, of related fossil resins, and of modern resins has utilized infrared (IR) spectroscopy (*2*), gas chromatography with mass spectrometry (GC/MS) (*3*), and solid state nuclear magnetic resonance (NMR) spectroscopy (*4*). All three methods have provided critical information in understanding the structure of resins and in distinguishing them on a geographical basis. It is the latter consideration that is important for archaeology. We have used the NMR method because it examines the entire, bulk material, whereas GC/MS, even with pyrolysis methods, samples only part of the resin. NMR has proved to be diagnostic for a much wider selection of materials than IR spectroscopy. NMR moreover has proved to be particularly useful in providing spectral distinctions between numerous different resins on a worldwide basis (*5*).

Although much of the study of fossilized resins has focused on materials of European origin, resins also are found on all continents (except Antarctica so far). In our study of modern resins (*5, 6*), we have found four major botanical sources with distinct NMR signatures: *Pinus*, *Hymenaea*, *Agathis*, and Burseraceae. In addition, there are other materials that are closely related to these genera, for example *Copaifera* to *Hymenaea* and *Araucaria* to *Agathis*. There are numerous other modern genera of resin-producing plants that contribute to a lesser extent, and we are in the process of characterizing many of these materials.

Fossilized resins also may be classified according to their NMR signatures. We (*5*) have designated four categories in this fashion (Anderson, Botto, et al. have provided a parallel classification based on GC/MS characteristics (*3*)). Our Group A is found throughout the world and has been traced forward to the modern conifer genus *Agathis* (Anderson's Class Ib in part). Our Group B also is found throughout the world but has not been associated with a modern genus. It may correspond to Anderson's Class II. Our Group C comprises amber of the Baltic Sea and related materials. It is part of Anderson's Class Ia but has clearly distinct NMR characteristics. Although it has a coniferous source, paleobotanists have not agreed on a family or genus. Our Group D is closely related to modern *Hymenaea* and corresponds to Anderson's Class Ic. Group D differs from the other groups in that the plants were leguminous angiosperms rather than conifers. There may be other minor groupings, but these four have been found to represent the vast majority of fossilized resins studied to date.

In our present study, we are focusing on the fossilized and modern resins of Africa and the Americas, which have been characterized much less than those of Europe. We have published partial studies in the past, including analysis of selected samples from a few sites in Africa, South America (*7*), North America (*8*), the Dominican Republic (*9*), and Mexico (*10*).

Examination of these materials is interesting not only from archaeological and botanical perspectives, but also from a geological point of view. Africa and South America were connected from approximately the Upper Devonian (ca. 360 Ma) to approximately the Upper Jurassic-Early Cretaceous (150-100 Ma) (*11-13*), well after the origin and expansion of the angiosperms during or before the Cretaceous (ca. 140-190 Ma) (*12, 14, 15*). North and South America were connected through a dried isthmus of Panama only from about the Early Pliocene (ca. 3 Ma) to the present, although broad connections of the continental shelves are believed to have existed earlier (*12*). Thus plant evolution could have resulted in greater similarity between resinous plants of Africa and South America than between those of the North and South American landmasses. Numerous examples of African-South American biogeographical affinities have been described (*16, 17*). We now have expanded our study of resins from these regions by solid state NMR spectroscopy and report those results herein.

## Results

We have obtained new resin samples from Africa (Mauritania, Madagascar, Congo, Morocco), South America (Guyana, Uruguay, Brazil), North America (North Carolina, North Dakota), the Caribbean (Puerto Rico), and Europe (Spain). The samples were provided by the following sources: Mauritania from the Gem and Mineral Show (Tucson, AZ); Congo from G. O. Poinar, Jr.; Morocco from the Gem and Mineral Show (Tucson, AZ); Madagascar (carbon-dated to about 125 years old) from T. Kapitany and J. Fell (the Cap d'Ambre region and south of there in northern Madagascar); Guyana (carbon dated to about 400 years old) from A. J. Lewis, Jewel of the Rainforest; Uruguay from C. Triad; Brazil (Matto Grosso) from N. E. de Oliveira, Mosaico Mineral Art; North Carolina from G. W. Powell, Jr., of Falls Church, VA; North Dakota (southwestern corner between Marmath and Rhame, approximately 65-67 Ma) from J. Ladin; and Spain (Alava, Early Cretaceous, approximately 100-140 Ma) from C. Vigil, Vitoria, Spain.

Samples collected from nature include the very dark brown to black resin of *Guiacum officinale* (Zygophyllaceae), known locally as guayacán, whose resin tends to be spherical, collected by M. Canals in Guánica (Puerto Rico) and the whitish, malleable, and the highly odorous resin of *Dacryodes excelsa* (Burseraceae), known locally as tabonuco, collected by F. Wadsworth and one of the authors (J. A. S.-B.) in Río Grande, Puerto Rico. All samples were obtained with appropriate permits.

Samples were ground in a mortar to a fine powder and loaded into a Doty Zirconia rotor, sealed with Aurum caps. About 100 mg of material was required. Samples were spun typically at 7 kHz in a Varian 300 MHz spectrometer operating at 75.413 MHz for $^{13}C$. Spectra were obtained only for $^{13}C$ with a 4.9 μs pulse, a 2200 μs contact time, a 5 s repetition delay, a 150 ms acquisition time, a 30 kHz spectral window, about 256 transients, and, for interrupted decoupling, a 50 μs delay. Two spectral modes were recorded for each sample, with normal decoupling and with interrupted decoupling. Adamantane peaks at δ 38.3 and 29.2 served as the external reference, and chemical shifts were converted to tetramethylsilane at δ 0.0.

## Discussion

Our past work on African and American resins has found Group A materials (*Agathis*-like) in several North American locations, including Alaska, Alberta, Manitoba, Kansas, Mississippi, and New Jersey. On the other hand, only a single site in Arkansas produced materials from Group B. All other Group B samples came from Asia and the Pacific. No Baltic-like samples (Group C) have been found in either Africa or the Americas. The *Hymenaea*-like samples of Group D previously were found in the Dominican Republic, the Mexican state of Chiapas, Colombia, Kenya, Tanzania, Congo, and Madagascar. With new analyses reported herein, we have been able to expand the geographical scope of all these groups except Group C, which retains its singular association with the Baltic region.

The Mauritanian copal (Figure 1) has the characteristic spectra of *Hymenaea* (see Figure 2 of ref. 6). With interrupted decoupling there are the four diagnostic peaks in the saturated region (δ 10-50). With normal decoupling the exomethylene resonances are clearly visible at δ 108 and 148. The spectra of Congo copal are essentially identical to those of *Hymenaea*, although the peak at δ 33 is doubled, a phenomenon not previously noted for this grouping. We cannot precisely define the botanical classification, so that minor differences might arise if the material for example is *Copaifera* rather than *Hymenaea*. Spectra of two new copal samples from Madagascar confirm that this material, previously uncertain based on a single sample (*6*), falls solidly into this grouping. From South America, Guyana copal exhibits the same spectral characteristics. The very small peak at δ 16 in the standard *Hymenaea* spectrum (*6*) is somewhat larger in this sample. The samples from Uruguay and Brazil are classic exam-

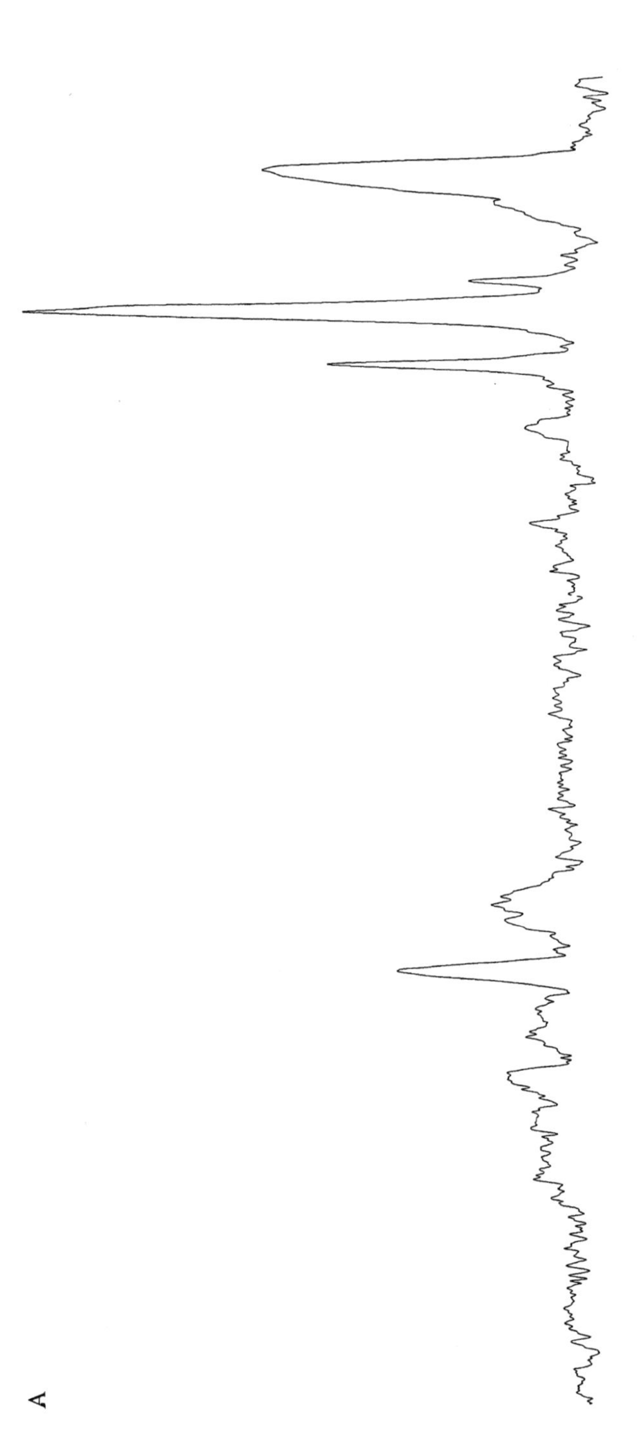

A

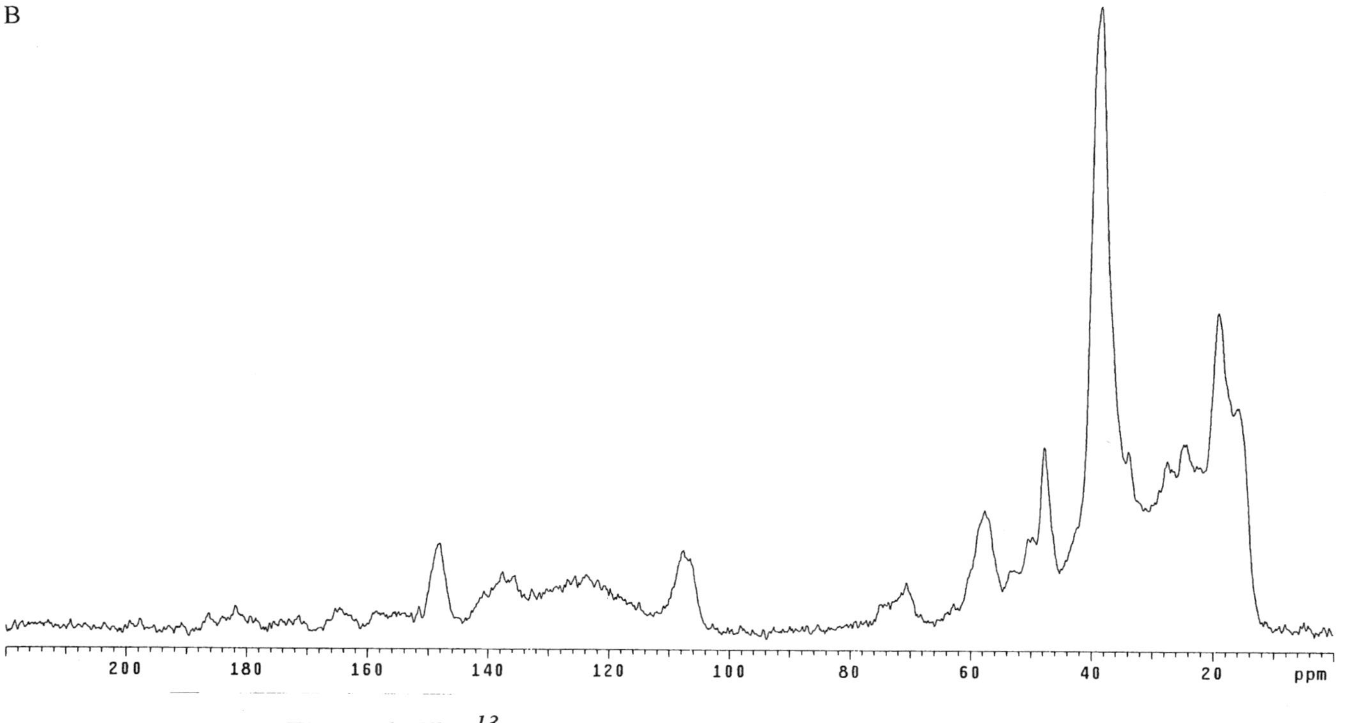


*Figure 1. The $^{13}C$ spectrum of copal from Mauritania, with normal (lower) and interrupted decoupling.*

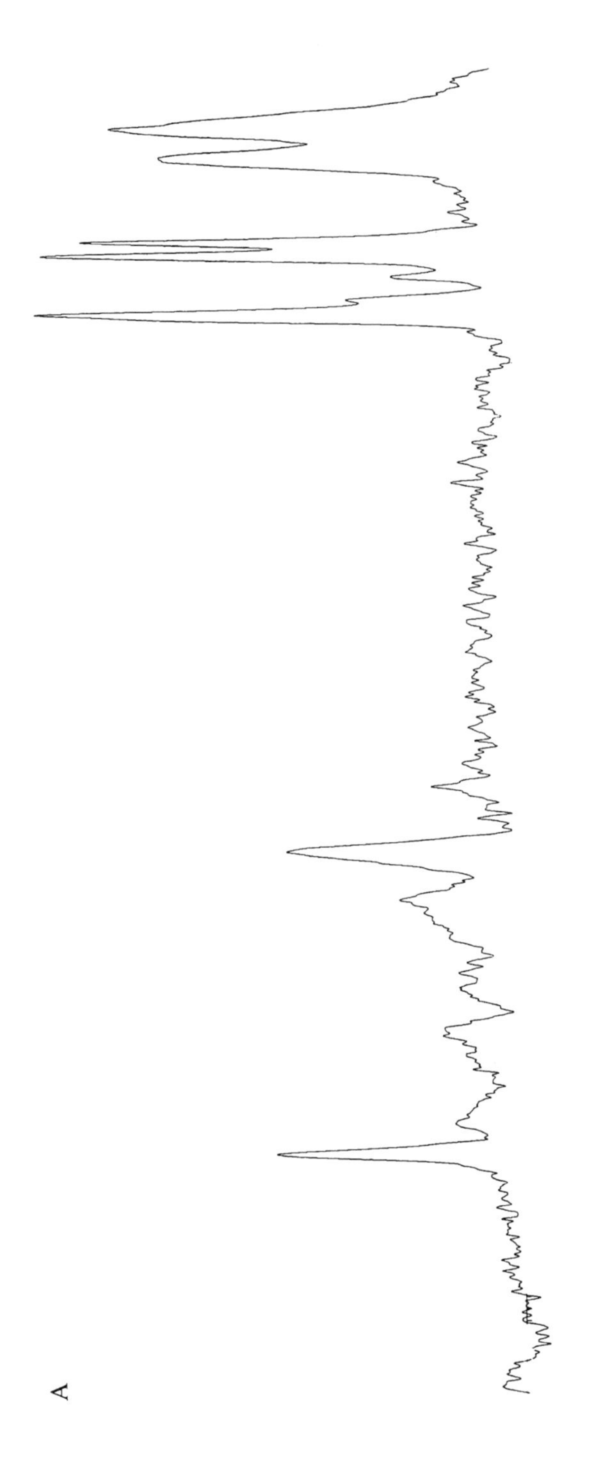
A

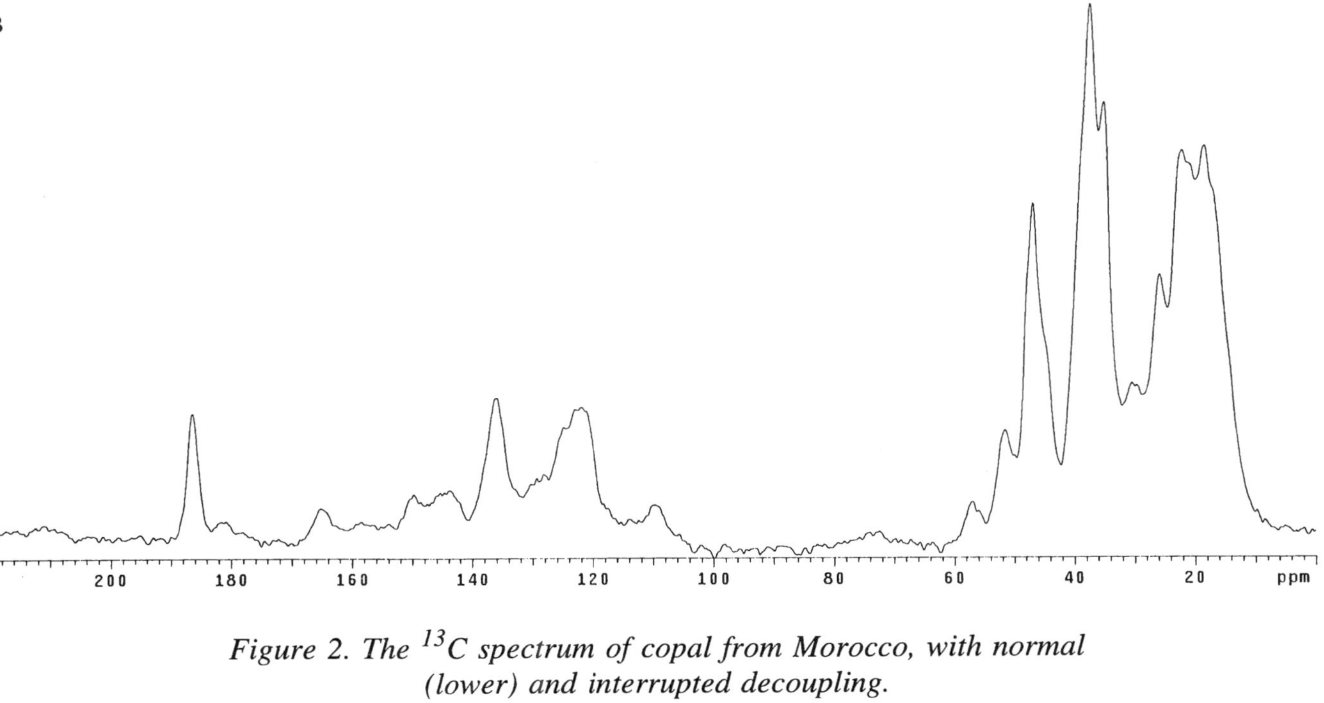


*Figure 2. The $^{13}C$ spectrum of copal from Morocco, with normal (lower) and interrupted decoupling.*

ples of the *Hymenaea* grouping. All of these samples show little evidence of decomposition in the spectra, indicative of recent origin, in contrast for example to the spectra we obtained from clearly older (more mature) samples from the Dominican Republic (*9*) and from the Mexican state of Chiapas (*10*).

The spectra of the sample of Moroccan copal (Figure 2) is very similar to those of the older, more altered (fossilized) *Hymenaea*-like samples from the Dominican Republic, Chiapas, and Tanzania (Group D) (*6*). Exomethylene resonances are present (in contrast to Groups A and B) but are decidedly reduced in intensity from those in recent material (Figure 1). With interrupted decoupling, the spectrum closely resembles those of modern *Hymenaea* and of the fossilized samples just listed. There are five peaks in the saturated region, as is found normally in the fossilized but not in the recent resins. The spectra of the sample from the new Chiapas site of La Primavera are very similar to those previously studied from Simojovel and Totolapa in that state.

Our Group B originally was very surprising, since it was not represented in Anderson's classification and it first contained samples only from Australia, New Guinea, and Arkansas, certainly an unorthodox combination. The later addition of the well known Burmese material burmite and of samples from India, Borneo, and Sumatra, however, documented a more widespread grouping (*6*), which may correspond to Anderson's Group II (Utah and Indonesia). We now have found that two samples from North Carolina also fall into Group B (Figure 3). These are from the town of Aurora. There are essentially no differences from the spectra of other members of this group (Figure 5 in ref. 6). In particular, there are no exomethylene carbons with normal decoupling, and there are three strong peaks in the saturated region with interrupted decoupling. This observation extends Group B to a completely different area of North America.

Interestingly, another sample, also from Aurora, North Carolina, did not fall into this category (Figure 4). Rather, it is clearly in the *Agathis*-related Group A with worldwide occurrence (*6*). The saturated region with normal decoupling is very characteristic of this group, but in addition there is the complete absence of exomethylene carbons with normal decoupling and the diagnostic pattern in the saturation region with interrupted decoupling. Thus even materials that are geographically very close can prove to have distinct paleobotanical origins, but the NMR spectra easily provide the distinction (compare Figures 3 and 4). This sample from North Dakota also falls into Group A, aligning itself with materials previously analyzed from Alberta and Manitoba. The sample from Spain is very similar, aligning itself with materials previously analyzed from France and Switzerland, for example.

Modern resin from Río Grande, Puerto Rico, identified as *Dacryodes excelsa* (family Burseraceae) and called tabonuco, gives spectra (Figure 5) that are identical in almost all details to the published spectra of the Burseraceae resin, probably of genus *Bursera* from Mexico (Figure 2 in ref. 6). This observation establishes the chemical identity of these two materials. As Burseraceae resins are triterpenoid (in contrast to diterpenoid *Agathis* and *Hymenaea*), we conclude that *Dacryodes* resin also is triterpenoid.

Finally, we present an entirely new spectrum of resin from Guánica, Puerto Rico, identified as *Guiacum officinale* (family Zygophyllaceae) and called guayacán (Figure 6). This remarkable spectrum contains enormous unsaturated resonances, between δ 110 and 150. We have not seen such strong resonances in this region of the spectrum of any other resins. This region contains both alkenic and aromatic resonances. The saturated region is relatively weak from δ 10 to 50 but exhibits a very strong peak at δ 60, usually characteristic of C-O (not the anomeric carbon of sugars, O-C-O, which resonates at higher frequency, close to δ 100, and indicates gum resins in this context; the small peaks at ca. δ 90 may fall into this category). Consequently, the spectra suggest a possible phenolic resin polymer containing the functionality Ar-O-C. With interrupted decoupling all peaks disappear except the three at δ 58, 135, and 146.

## Conclusions

Fossilized resins only rarely can be attributed to a particular botanical grouping, and modern resins can be so identified only when they are harvested from or under the producing tree. Modern resins or copals obtained from collectors, dealers, museums, or archaeological sites are of undetermined botanical origin unless specifically harvested by the collector, dealer, or curator in close relationship to the producing plant. Thus NMR provides a useful method for botanical and paleobotanical assignment. Expansion of our knowledge of resin classes enables the NMR spectroscopist to provide more reliable assignments of archaeological resins to their original geographical source.

We now have expanded almost all the groups of resins defined in our recent article (*5*). Group D is derived from *Hymenaea* and related leguminous angiosperms. The geographical scope of recent materials previously included the sites of the Dominican Republic, Colombia, Kenya, and Congo. The scope of fossilized sites included the Dominican Republic, the sites of Simojovel and Totolapa in the Mexican state of Chiapas, Tanzania, and possibly Madagascar in Africa. We now have extended the scope of the modern materials to include the East African country of Mauritania and the South American countries of Guyana,

A

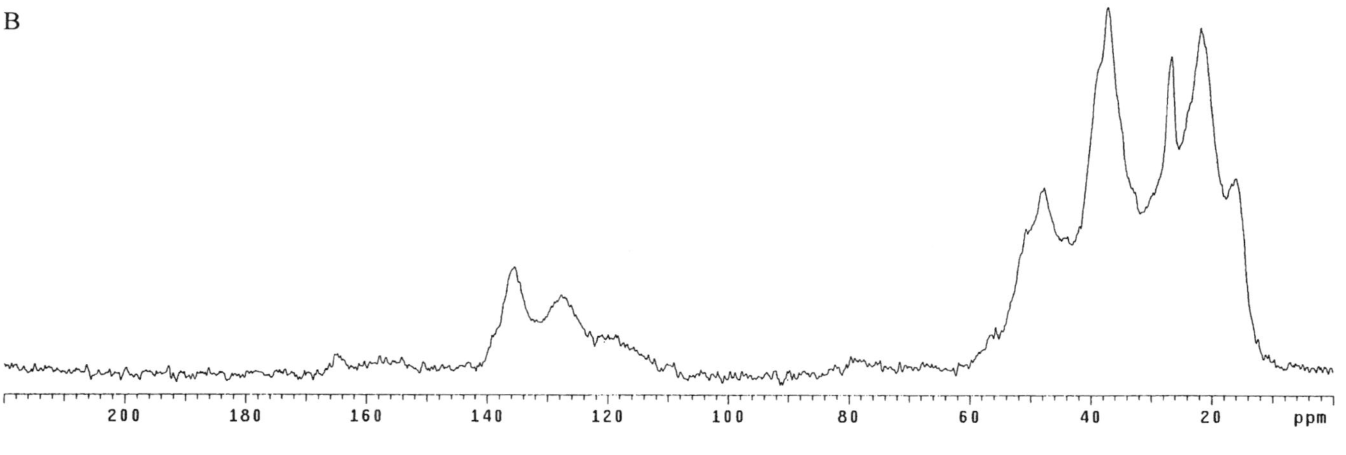


*Figure 3. The $^{13}C$ spectrum of fossilized resin from Aurora, North Carolina, with normal (lower) and interrupted decoupling.*

A

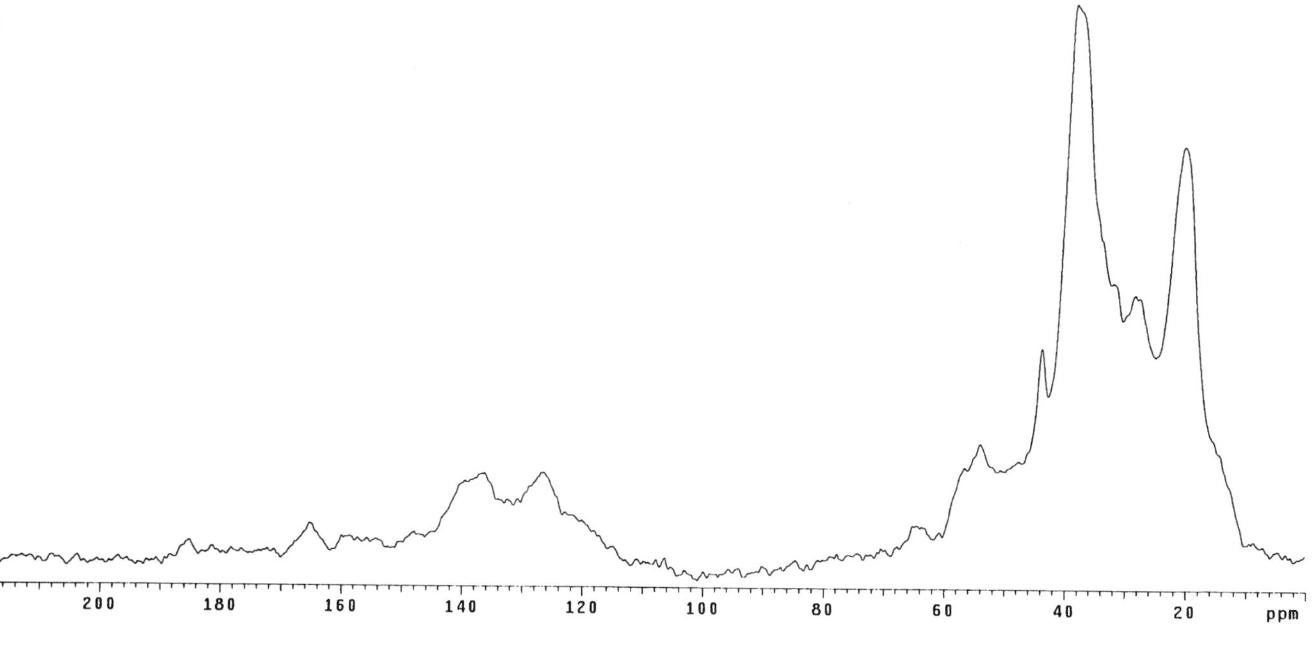


*Figure 4. The $^{13}C$ spectrum of fossilized resin from Aurora, North Carolina, with normal (lower) and interrupted decoupling. This sample is different from that in Figure 3.*

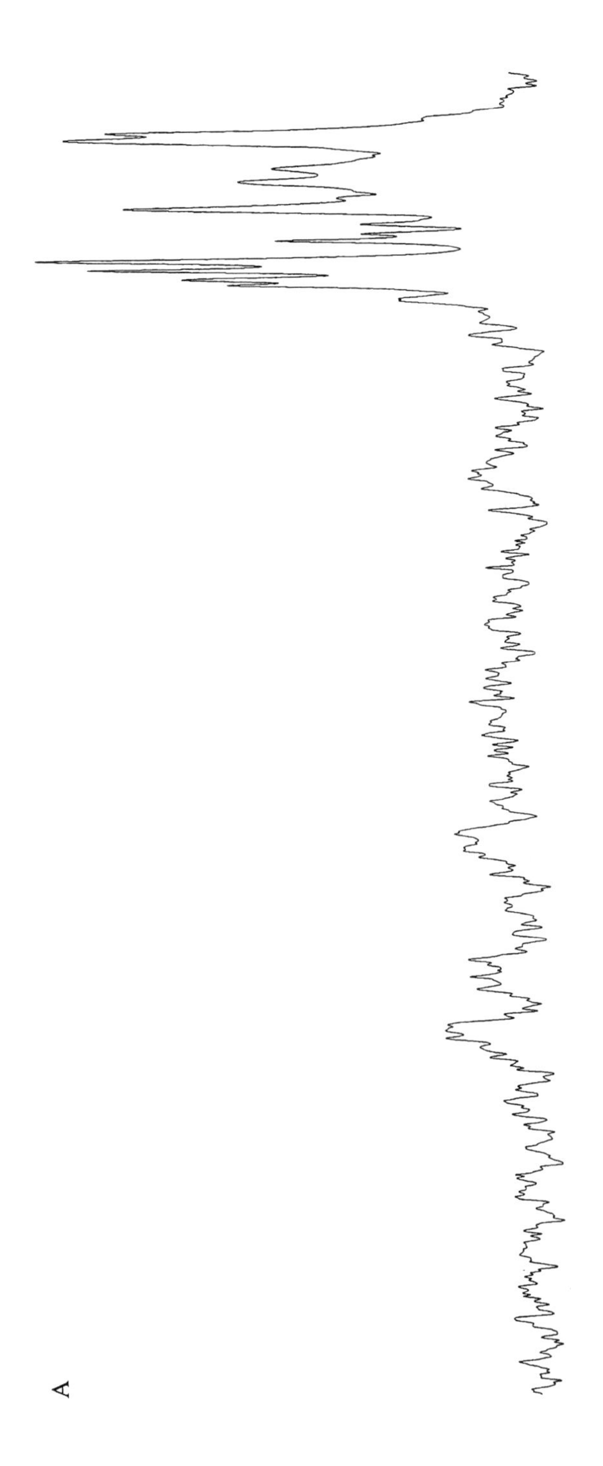
A

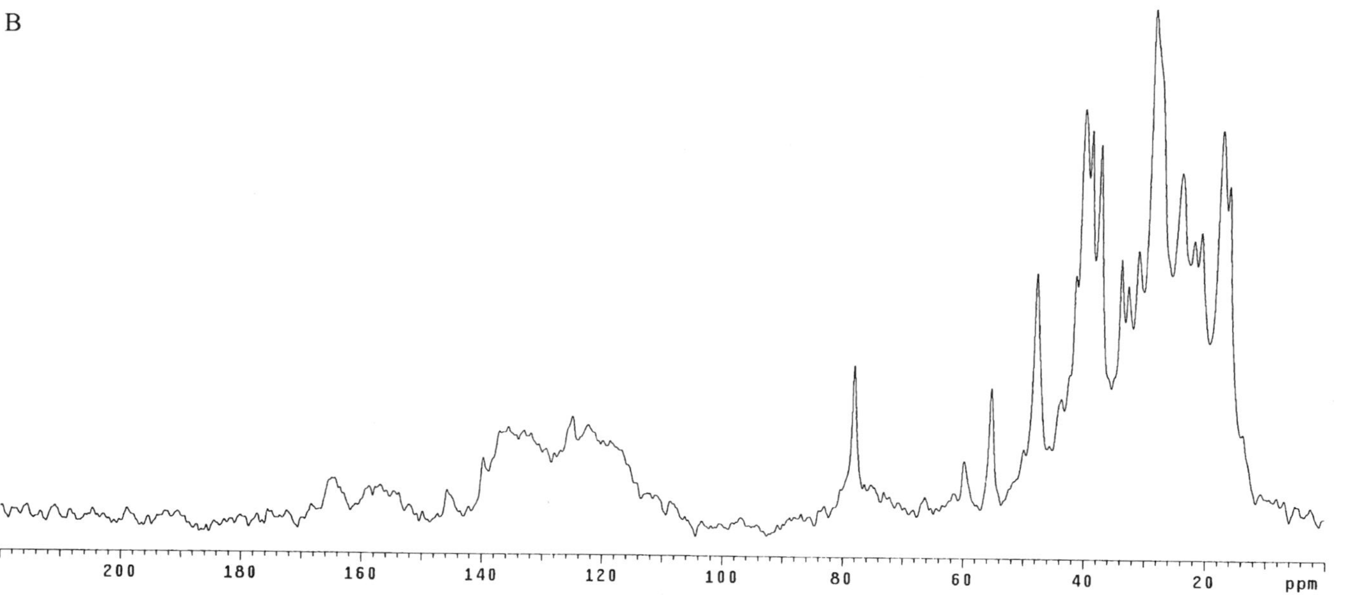


*Figure 5. The modern resin Dacryodes excelsa from Río Grande, Puerto Rico, with normal (lower) and interrupted decoupling.*

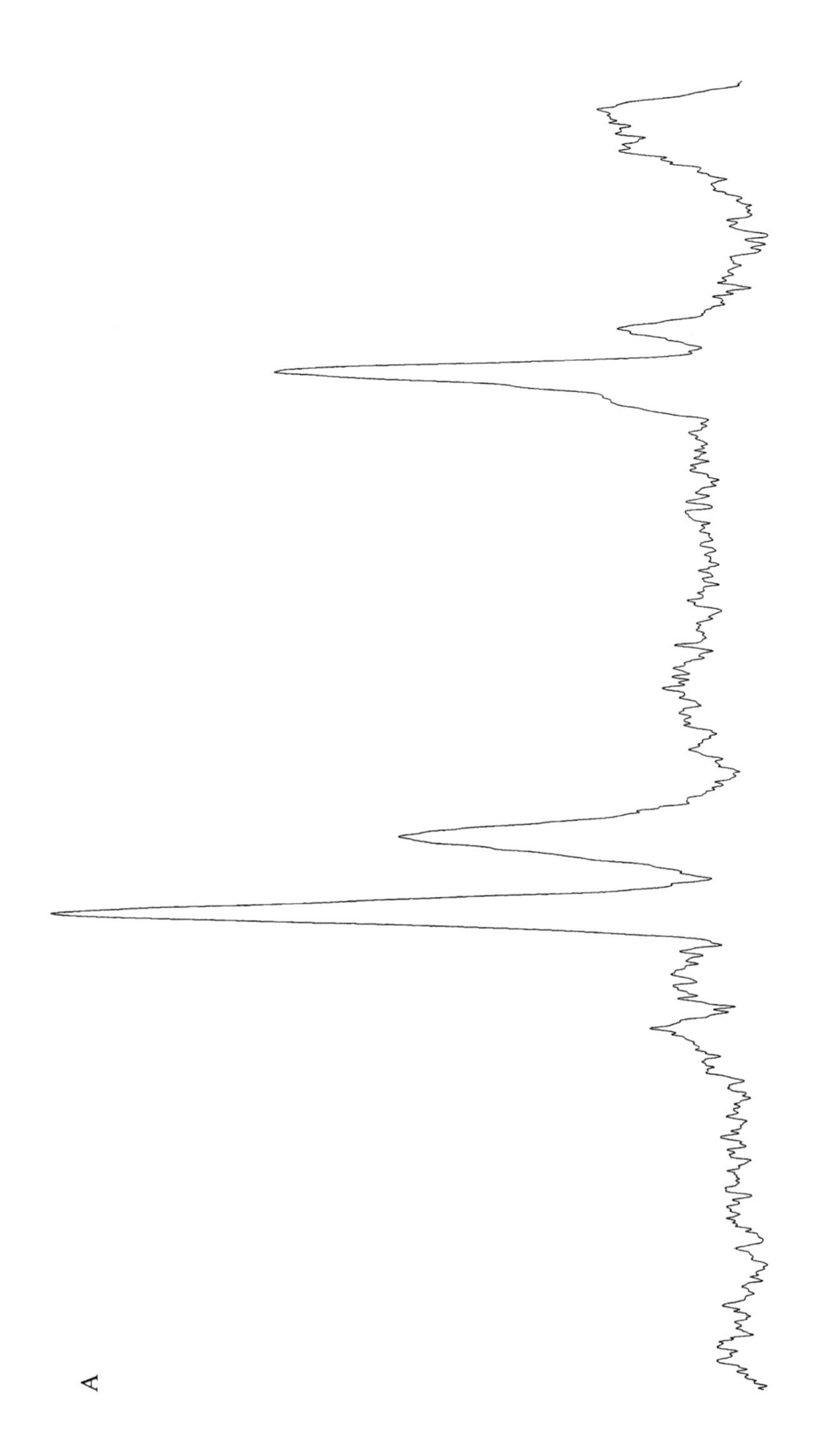
A

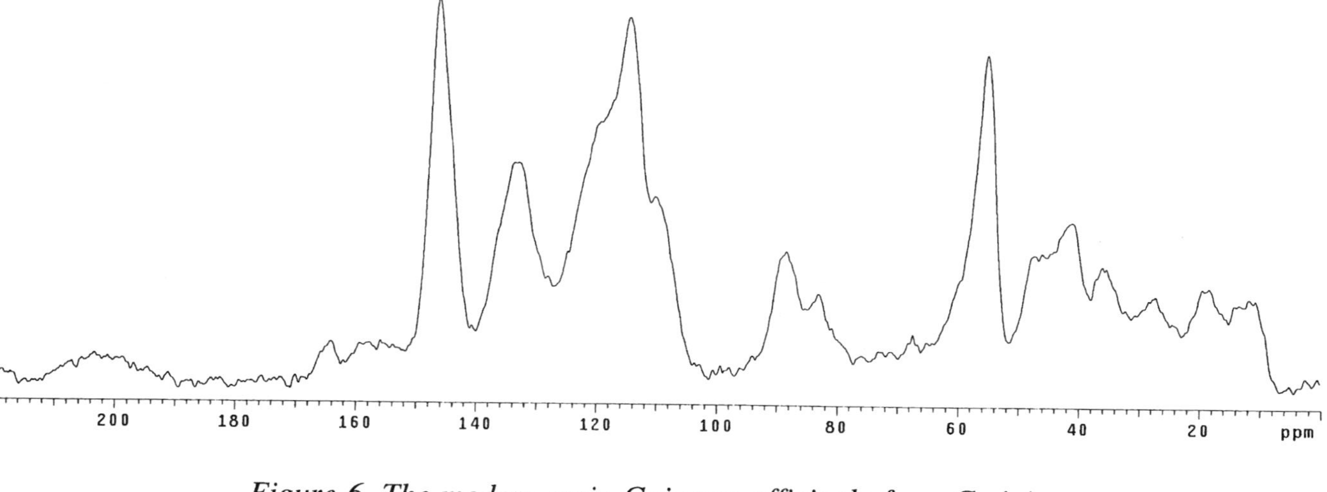


*Figure 6. The modern resin Guiacum officinale from Guánica, Puerto Rico, with normal (lower) and interrupted decoupling.*

Uruguay, and Brazil. With inclusion of Colombia, Group D modern samples now derive from a large part of that continent. The scope of fossilized Group D samples now includes the North African site of Morocco and an additional Chiapas site called La Primavera. The chief spectral distinction between modern and fossilized materials is the strong exomethylene resonances in the former and the reduced exomethylene resonances in the latter.

Our Group B extends almost around the world, previously including samples from India, Burma, Borneo, Sumatra, Papua New Guinea, Australia, and Arkansas in the United States. We now report a sample from North Carolina in the United States that exhibits the classic pattern of this group, so that Group B no longer is an outlier in the North American context. We have not associated this group with a specific plant source, but botanical studies on material from Arkansas and Sumatra have suggested that it may be the family Dipterocarpaceae.

The *Agathis*-related Group A is the most widespread, represented extensively from Europe, Asia, Australia, and North America. We now have found additional sites in Spain, North Dakota, and North Carolina. The North Carolina site is of particular interest, because the material was found in the same town as the new example of Group B. The materials, however, had distinct and characteristic spectra. Thus one can make no assumptions about classification of new finds without spectral examination.

No expansion of Group C, Baltic amber, has been made. It remains a unique material associated with the countries around the Baltic Sea and possessing very diagnostic NMR and IR spectra.

Undoubtedly there are many modern resins, still unstudied, that may provide distinct and characteristic NMR spectra. As we describe herein, *Dacryodes excelsa* from Puerto Rico exhibits the same spectra as materials from the Burseraceae family from Mexico. On the other hand, a sample of resin from *Guiacum officinale* produces a unique new NMR spectrum.

It is clear that studies of modern and fossilized resins will continue to expand the existing groupings but may at the same time uncover new categories and refine the worldwide classification of resins.

## Acknowledgments

We thank the numerous collectors mentioned above, who have donated their time and resources to provide modern and ancient samples. This research was partially supported through a Mednick Memorial fellowship from the Virginia Foundation of Independent Colleges (Richmond, VA) to author J. A. S.-B.

## References


1. Mills, J. S.; White, R. *Organic Chemistry of Museum Objects* 2nd ed.; Butterworth Heinemann: Oxford, UK, 1994; Chapter 8. Rice, P. C. *Amber The Golden Gem of the Ages*; Van Nostrand Reinhold: New York, NY, 1980. Grimaldi, D. A. *Amber: Window to the Past*; Harry N. Abrams: New York, NY, 1996. Fraquet, H. *Amber*; Butterworths: London, UK, 1987. Ross, A. *Amber*; Harvard University Press: Cambridge, MA, 1999. Poinar, G.; Poinar, R. *The Quest for Life in Amber*; Addison Wesley: Reading, MA, 1994. Poinar, G. O.; Poinar, R. *The Amber Forest*; Princeton University Press: Princeton, NJ, 1999.
2. Beck, C. W. *Appl. Spectrosc. Rev.* **1986**, *22*, 57-110.
3. Anderson, K. B.; Winans, R. E.; Botto, R. E. *Org. Geochem.* **1992**, *18*, 829-841. Shedrinsky, A. M.; Grimaldi, D. A.; Boon, J. J.; Baer, N. S. *J. Anal. Appl. Pyrolysis*, **1993**, *25*, 77-95.
4. Lambert, J. B.; Frye, J. S. *Science* **1982**, *217*, 55-57.
5. Lambert, J. B.; Poinar, G. O., Jr. *Acc. Chem. Res.* **2002**, *35*, 000-000.
6. Lambert, J. B.; Shawl, C. E.; Poinar, G. O., Jr.; Santiago-Blay, J. A. *Biorg. Chem.* **1999**, *27*, 409-433.
7. Lambert, J. B.; Johnson, S. C.; Poinar, G. O., Jr. In *Amber, Resinite, and Fossil Resins*; Anderson, K. B.; Crelling, J. C., Eds.; ACS Symposium Series 617; American Chemical Society: Washington, DC, 1995; pp 193-202.
8. Lambert, J. B.; Frye, J. S.; Poinar, G. O. *Geoarchaeology* **1990**, *5*, 43-52.
9. Lambert, J. B.; Frye, J. S.; Poinar, G. O., Jr. *Archaeometry* **1985**, *27*, 43-51.
10. Lambert, J. B.; Frye, J. S.; Lee, T. A., Jr.; Welch, C. J.; Poinar, G. O., Jr. In *Archaeological Chemistry IV*; Allen, R. O., Ed.; Advances in Chemistry Series 220; American Chemical Society, Washington, DC, 1989; pp 381-388.
11. *Tropical Forest Ecosystems in Africa and South America: A Comparative Review*; Meggers, B. J.; Ayensu, E. S.; Duckworth, W. D., Eds.; Smithsonian Institution Press: Washington, DC, 1973.
12. *The evolution of terrestrial ecosystems consortium*; Behrensmeyer, A. K.; Damuth, J. D.; DiMichele, W. A.; Potts, R.; Sues, H-D.; Wing, S. L., Eds.; The University of Chicago Press: Chicago, IL, 1992.
13. White, M. E. *The Flowering of Gondwana. The 400 Million Year Story of Australia's Plants*; Princeton University Press: Princeton, NJ, 1990.
14. DiMichele, W. A.; Stein, W. E.; Bateman, R. M. In *Evolutionary Paleo-ecology. The ecological context of macroevolutionary change*; Allmon, W. D.; Bottjer, D. J., Eds.; Columbia University Press: New York, NY, 2001, Chapter 11.
15. Sanderson, M. J.; Doyle, J. A. *Amer. J. Botany* **2001**, *88*, 1499-1516.
16. Huber, F. M.; Langenheim. J. L. *Geotimes* **1986**, *31*, 8-10.
17. Poinar, G. O., Jr. *Experientia* **1991**, *47*, 1075-1082.

## Chapter 7

# Characterization and Radiocarbon Dating of Archaeological Resins from Southeast Asia

**C. D. Lampert[1], I. C. Glover[2], C. P. Heron[1], B. Stern[1], R. Shoocongdej[3], and G. B. Thompson[1]**

**[1]Department of Archaeological Sciences, University of Bradford, Bradford BD7 1DP, United Kingdom**
**[2]Institute of Archaeology, University College London, London WC1H 0PY, United Kingdom**
**[3]Department of Archaeology, Silpakorn University, Bangkok-10200, Thailand**

Archaeologists have identified evidence for the use of resins in association with some Southeast Asian ceramics. Resins are largely mixtures of terpenoids with similar molecular structure and numerous isomers. Well-established methods to investigate solvent soluble components of modern resins include gas chromatography and GC-mass spectrometry, sometimes combined with pyrolytic techniques to explore the high MW or polymeric insoluble fraction. This approach is equally applicable to the characterisation of archaeological resins. Although resins are a widespread natural resource in Southeast Asia, it is not clear which of the many resinous species were exploited in the past. To this end, a study of archaeological resins from Southeast Asia is underway with objectives of identification, examination of use, comparisons across geographical distance and time, and evaluating their merit for radiocarbon dating. Samples have been analysed using GC and GC-MS and preliminary results suggest that Dipterocarp resins were used as adhesives or sealants.

# Background

Natural resins are found throughout the world (Figure 1) and their many valuable physical and chemical properties have led to their exploitation for a broad variety of purposes, both in the past and the present. Amongst other things they are flammable, insoluble in water, adhesive, aromatic and can act as biocides. Traditionally, resins have been used as adhesives and varnishes, as components of pigment media, as sealants and waterproofing agents, as incense or in perfumery, as components of torches or firelighters and for a diversity of medicinal purposes. Gianno (*1*) provides a detailed evaluation of applications resins have found worldwide.

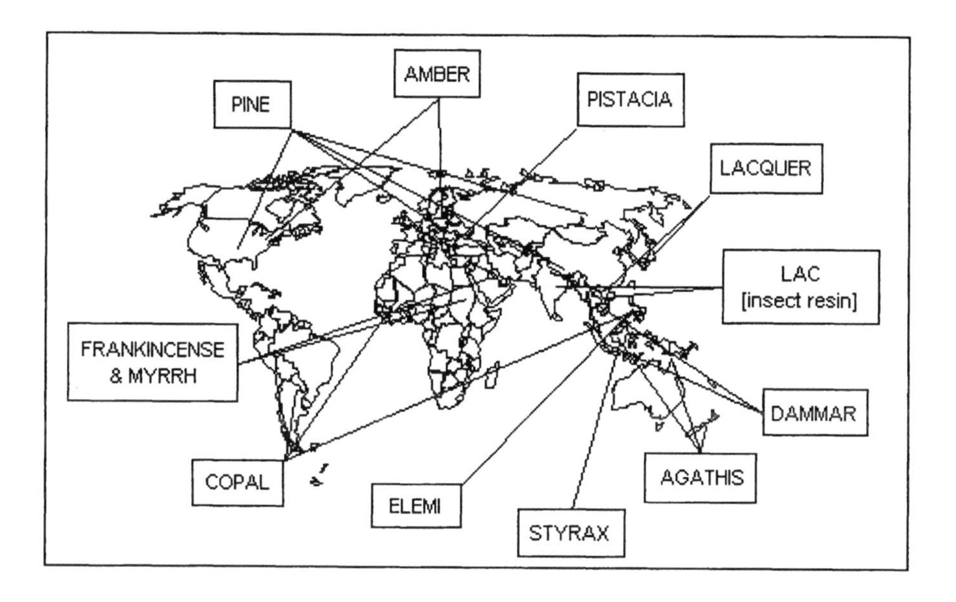


*Figure 1. Some of the many resins found worldwide.*

Evidence for the use of resins in association with ceramics has been identified at a number of archaeological sites within Southeast Asia. Resins may be found in close association with an inorganic substrate such as a ceramic vessel due to their deliberate application, for example in the form of a coating, varnish or adhesive.

Alternatively, they may be deposited during use, possibly as one component of a combination of substances mixed or contained in a pot or from the resins themselves having been stored or traded in a vessel. Deliberate application, particularly where a resin is used as a coating, may be intended to alter the properties of a vessel, for example to modify its permeability (*2*) or heating effectiveness (*3*), although a resin coat might also be applied for purely aesthetic reasons.

One aspect of this research can be summarised as the identification of resinous resources used at specific sites in prehistory, an examination of how they were used and a comparison of the use of resins across geographical distance and time. Studying resins found in association with archaeological ceramics has the potential to further our understanding of the acquisition and use of natural organic resources by prehistoric populations and may shed light on the potential movement of resins as a traded resource. It may add to our awareness of technologies exploited in prehistory or broaden our understanding of vessel function.

The other major objective is to explore the potential of resins for radiocarbon dating. Initially, we must consider whether it is possible to obtain valid, reliable and repeatable radiocarbon dates from resins. Once it can be established whether the radiocarbon dates we obtain are dependable, the question of whether they can be used to directly date the ceramics with which they are associated can be addressed.

## Resins in Southeast Asia

The flora of Southeast Asia, a geographical region stretching across some 4000 miles from the borders of Myanmar to Papua New Guinea, is diverse and home to many species of plants. Many produce exudates, including resins, often in response to injury. A large number of plants in Southeast Asia could have been used as a source of resin by ancient peoples. Of these, plants from the Dipterocarpaceae and Pinaceae families are widespread and perhaps the most plausible sources for resins used in association with ceramics, although plants from the Sapotaceae, Burseraceae and other families should not be ruled out. A smaller number of resins produced by insects are also found in the region but this paper concentrates on plant resins.

As might be expected over such a broad area, significant floristic variations exist. The main differences are between the flora present on the bulk of mainland Southeast Asia and that of the Malay Peninsula and the islands to the east. Altitude and climate, in particular levels of rainfall in a given area, largely dictate the local flora. Each region's flora is distinct, without a gradual merging between adjacent regions (*4*), and each of the forest types can offer numerous resinous species.

Mainland Southeast Asia is seasonal and is comparatively dry. Deciduous dipterocarp forests are widespread (*5*), with conifer forests at higher altitudes (*6*). In comparison, areas to the south and east, collectively known as Malesia, are wetter with evergreen tropical rainforests, again rich in dipterocarp species. However, there is a drier, more seasonal band with a high proportion of conifers covering the central zone of the Philippines, Sulawesi, the Moluccas and into Java before tropical rainforest takes over again in Irian Jaya and the east (*4*).

## Existing evidence for the use of resins in Southeast Asia

Ethnographic and archaeological evidence pointing to the association of resins and ceramics in Southeast Asia exists from both the mainland and insular regions. Resin-coated potsherds were excavated during the 1960s at Spirit Cave, in Northwestern Thailand, a Hoabinhian rock-shelter site and potsherds and ceramic spindle whorls bearing traces of a coating of a resinous substance were more recently recovered from burials at the Iron Age site of Noen U-Loke on the Khorat Plateau in Northeastern Thailand. A coating, thought to be a resin, has been observed on the interior of ceramic vessels from the Plain of Jars in Laos and a resinous deposit, possibly used as an adhesive, was discovered around the rims of lidded funerary urns at Hau Xa, a late Sa Huynh coastal site in Central Vietnam. The use of resins on the mainland is not restricted to prehistory. Resins, thought to have been used as binders in brickwork and mortars, were used in the construction of some Cham temples in Vietnam. In addition, evidence exists for trade in certain resins. Ceramic transport vessels were found on 15$^{th}$-17$^{th}$ century shipwrecks off the coast of Thailand with their cargo of resin still intact and, on these same ships, resins were also used to seal the lids onto large storage jars (*7*).

Examples of resin-coated ceramics from archaeological contexts in insular Southeast Asia include potsherds from a number of sites in Malaysia and Sumatra (*8*). Ethnographic examples of an organic post-firing treatment come from Moluku, Sulawesi and from Northwestern Luzon, where pots hot from firing are coated with a pine resin (*Agathis philippinensis*) and it has been shown that the location of the resin on the pot is dependent on the proposed use of the vessel (*9*). Similarly coated pots are manufactured in the Central

Moluccas (*10*) where *Agathis alba* resin is used and some potters in Papua New Guinea use a resin coat to seal water pots (*11*).

## Resin chemistry

The chemistry of resins is complex but they are related to plant essential oils and are largely terpenoid based. Terpenoids, sometimes also referred to as terpenes, are "usually cyclic, unsaturated hydrocarbons, with varying degrees of oxygenation in the substituent groups (alcohol, aldehyde, lactone, etc.) attached to the basic skeleton" (*12*). The cyclic carbon 'backbone' or 'skeleton' in terpenoids is of importance to studies of archaeological resins as, although functional groups may alter or be lost over time, the carbon ring-structure is generally preserved, and can be used as a biomarker to relate a degraded substance in archaeological material back to its equivalent compound in an authentic resin.

The terpenoids can be defined as "a group of natural products whose structure may be divided into isoprene units" (*13*). Isoprene, or isopentenyl, is a $C_5$ (five carbon atom) compound, which can be viewed as the terpenoid building block (Figure 2). Both the lower terpenoids (monoterpenoids and sesquiterpenoids) and the higher terpenoids (diterpenoids and triterpenoids) are built up from multiples of these isoprene units. When freshly exuded, resins tend to be liquid or semi-liquid, as they contain either one or the other of the two higher terpenoids in solution in a mixture of the more volatile lower terpenoids.

The terms 'terpene' and 'terpenoid' appear in many publications as largely interchangeable, although 'terpene' is sometimes used to refer only to hydrocarbons based on the $C_5$ isoprene unit whilst 'terpenoid' may encompass compounds which are derived from a number of $C_5$ isoprene units but have a varying number of carbon atoms, for example the $C_{27}$ steroids (*14*).

## Biosynthesis of resins

In practice, isoprene, although a convenient model to help understand the structural chemistry of terpenoids, is not the precursor to the terpenoid family in nature. In fact, the initial biosynthetic building block for the terpenoid family is mevalonic acid (*15*). From this compound, via a long series of intermediate steps, are formed an extensive range of terpenoid compounds (*16*), including those found in resins.

Many plants have biosynthetic pathways that can produce both mono- and sesquiterpenoids and these are often found together in plant tissues. However,

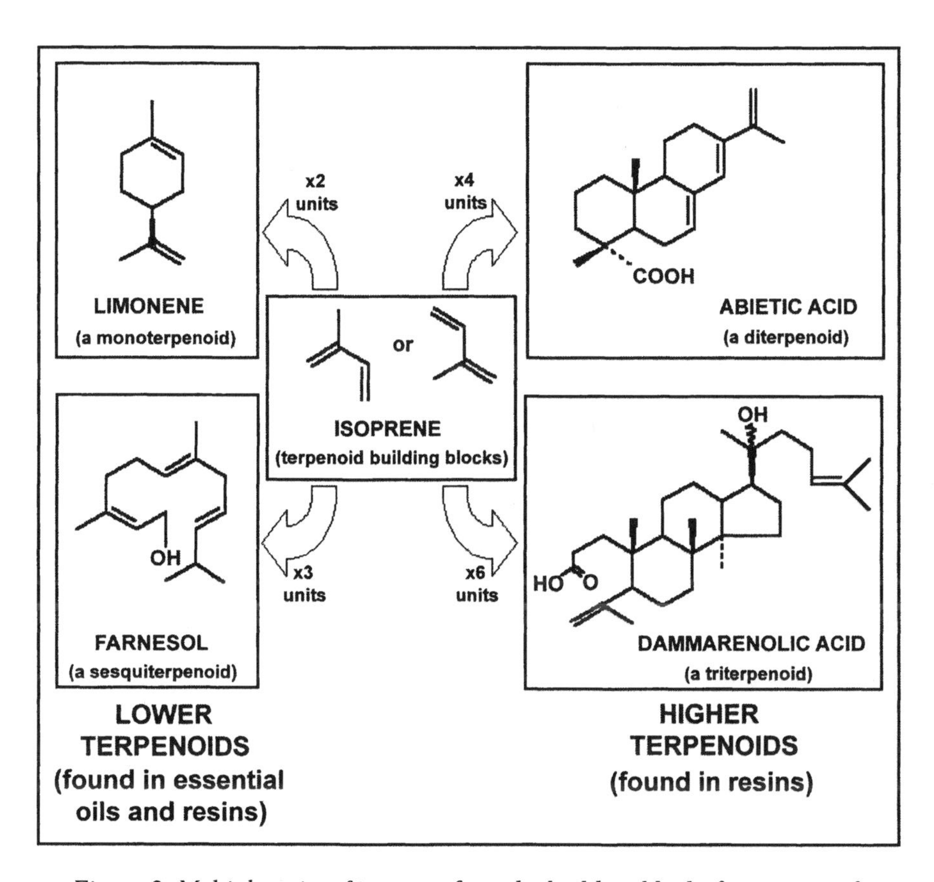


*Figure 2. Multiple units of isoprene form the building blocks for a range of terpenoids*

the genetic makeup of a given plant will permit it to biosynthesise either diterpenoid or triterpenoid compounds, but not both. This holds true for entire families of plants, and within these the sub-families, genera and species. Since di- and triterpenoids are not found together in the same plant, resins can be divided into two broad groups based on their higher terpenoid component. This forms a valuable first step in determining the botanical source of an unknown, for example archaeological, resin.

Mills and White (*17*) give an overview of the chemistry of many of the di- and triterpenoid resin producing plant families, showing some of the biomarker structures which can be used to identify the origin of different types of resin. For example, observation of diterpenoid compounds might suggest the plant that produced the resin might belong to the Coniferae or Leguminosae family. Conversely, determination of a triterpenoid component would rule out these families as the botanical source of the resin.

Within a plant family, there can be a significant divergence in the di- or triterpenoid compounds synthesised by each genus. Taking the Coniferae family as an example, the genera Araucariaceae, Cupressaceae and Pinaceae can be distinguished one from another by compositional differences in their diterpenoid biomarkers (*18*). Of the Dipterocarpaceae family, a number of genera have been differentiated by their chemical composition, but the majority of genera are said to be chemically homogeneous (*19*).

Whilst plant families and genera may be distinguishable by the presence of distinctive biomarkers, differentiation between species may be more difficult to achieve since two species from the same genus may share a very similar suite of biomarkers, differing only in the proportions of their components. This has been demonstrated by the similarities in chemical composition of samples of elemi resin from two species of *Canarium* (*20*), by similarities in the bled resins of several species of *Agathis* (*21*) and by the composition of a number of species of pine and other conifer resins tabulated in Mills and White (*17*). Care is necessary when variations in percentage composition are cited to identify specific species as noticeable variations can also occur between trees from the same species grown in different geographical locations (*22, 23*).

## Archaeological resins and changes in their chemistry

Due to their chemistry and physical properties, particularly their insolubility in water, the higher terpenoid components of resins are relatively stable and the preservation of resins in archaeological burial environments is not uncommon

(*24*). Indeed, so-called 'fossil' or 'semi-fossil' resins were once routinely collected for the 'dammar' trade by digging up masses of hardened resin from shallow soil deposits at the foot of old trees and true fossil resins are not infrequently found in soil or coal deposits.

However, resins found during archaeological excavations are unlikely to be entirely unchanged from their freshly exuded state and may be expected to have undergone some degree of chemical modification. A key difference between freshly exuded and archaeological resin is often the loss of many volatile lower terpenoids from the older material, although some of these can be preserved and may remain, particularly if trapped within a thick deposit or in bulk resin.

Primarily, it is the substituent or functional groups which are involved in any degradative reactions whilst the condensed ring-structure skeleton remains fundamentally unaffected, although some modifications due to breaking, formation or repositioning of C-C double bonds through hydrogenation and dehydrogenation may take place.

Even before deposition in the burial environment, a degree of chemical alteration may occur whilst an object is still in use. For example, where a thin film of resin is present as a coating or varnish, say on the exterior of a pot, some oxidative changes may be observed from exposure to light (*25*). Further oxidation is likely if the resin is subjected to heat. This may come about during technological processes to prepare the resin for use (e.g. preparation of a varnish). It may also take place during the application of a resin to a hot pot to form the coating or where a resin-coated vessel is heated, for example when used for cooking. In addition to the oxidative changes, some or all of the more volatile components may be driven off on heating.

Once in the burial environment, biodeterioration may promote further alterations in the structure of the higher terpenoids. Whilst the particular chemical character of the burial environment will dictate specific changes, once again, alterations likely to be encountered will include oxidation, hydrogenation, etc. Looking then at the composition of freshly exuded and aged resins, whilst the more stable terpenoids may be present in both, the older material is likely to exhibit a higher proportion of oxidised terpenoids.

# Methods

Samples of both modern resins and archaeological residues scraped from the surface of potsherds were solvent extracted using a 2:1 mixture of dichloromethane and methanol. A 3 ml aliquot of the solvent mixture was added to a weighed portion of sample material and sonicated for 3 minutes The samples were centrifuged at 20000 rpm for 5 minutes to settle any solid matter before a 1 ml aliquot was transferred to a clean vial. This was dried under a gentle stream of nitrogen gas prior to derivatization using ethereal diazomethane.

Gas chromatography was carried out using a Hewlett Packard 6890 gas chromatograph fitted with a Phenomenex ZB-1 capillary column (30 m length, 0.25 µm film thickness, 0.32 mm i.d.). A split ratio of 1:10 was used. The carrier gas was helium with a constant flow rate of 3ml/min. The injector was maintained at 300°C, the detector at 350°C. The oven was programmed to an initial temperature of 50°C for 2 minutes, increasing by 10°C per minute to 220°C then 5°C per minute to a maximum of 340°C, holding the maximum temperature for 12 minutes.

Gas chromatography-mass spectrometry was carried out using the column described above in a Hewlett Packard 5890 gas chromatograph series II with 5972 mass selective detector. The carrier gas was helium with a constant flow rate of 2ml/min and a head pressure of 4.6 psi at 50°C. The injector and interface were maintained at 300°C and 350°C respectively. The oven was programmed as above. The column was inserted directly into the ion source and electron impact (EI) spectra were obtained at 70eV with full scan from 50 to 700 *m/z*.

## Analysis of modern resins

A range of modern resin samples from the Dipterocarpaceae, Pinaceae and Burseraceae families (Table I), encompassing both diterpenoid and triterpenoid resins, have been examined using GC and GC-MS to act as reference materials against which to compare results from the archaeological samples. Wherever possible, the higher terpenoid peaks have been identified by comparison with previously published data.

**Table I. Modern reference materials**

| *Triterpenoid resins* | |
|---|---|
| *Family, genus & spp.* | *Common names* |
| Burseraceae | Manila elemi, white pitch, |
| *Canarium luzonicum* A. Gray [a] | occasionally dammar |
| Dipterocarpaceae | generally 'dammar' |
| *Balanocarpus heimii* King [a] | damar penak, |
| *Dipterocarpus alatus* Roxb [b] | minyak keruing, gurjun or wood oil, |
| *Dipterocarpus costatus* C.F. Gaertn [a] | sal |
| *Shorea robusta* C.F. Gaertn [a] | |
| *Diterpenoid resins* | |
| *Family, genus & spp.* | *Common names* |
| Araucariaceae | Manila copal, occasionally dammar |
| *Agathis alba* Jeffrey [a] | |
| *Agathis* spp. (probably *A. alba*) [b] | |
| Pinaceae | naval stores |
| *Pinus kesiya* Royle ex Gordon [a] | (yields rosin and turpentine) |
| *Pinus* spp. (probably *P. kesiya*) [b] | |
| *Pinus merkusii* Jungh & de Vries [a] | |

[a] courtesy of the Economic Botany collections, Royal Botanic Gardens, Kew

[b] collected in the field by I.C. Glover

## Triterpenoid resins from the Dipterocarpaceae family

The triterpenoid constituents of resins analysed are usually seen as a group of peaks eluting at around 30 minutes retention time under the analytical conditions used. Of the triterpenoid resins in Table I, four of the five come from the Dipterocarpaceae family. Their complexity and triterpenoid composition varies to some degree, but common to all are a number of peaks identifiable as compounds from the tetracyclic dammarane series (Figures 3-6). In addition to these, pentacyclic triterpenoids from the ursane and oleanane series are present. Although some peaks have yet to be conclusively identified, they are included since they are detected in more than one of the Dipterocarp reference samples and have also been observed in some archaeological material.

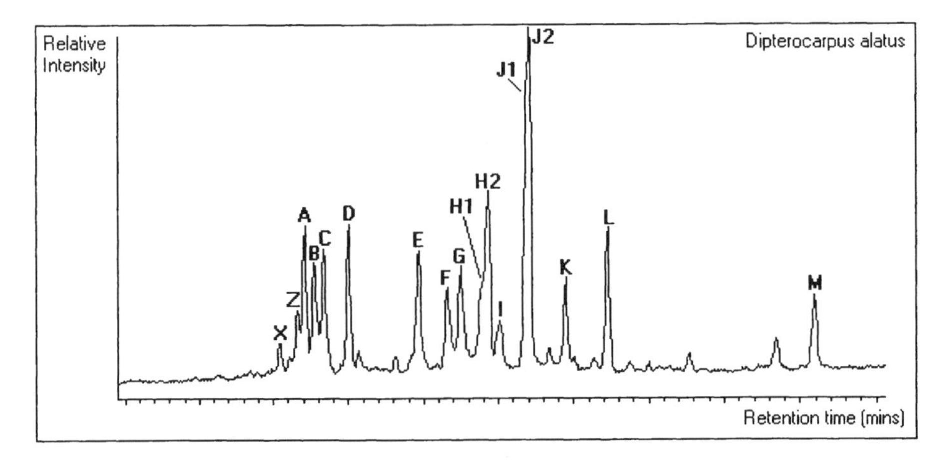


*Figure 3. Triterpenoid region of chromatogram of* Dipterocarpus alatus *resin*

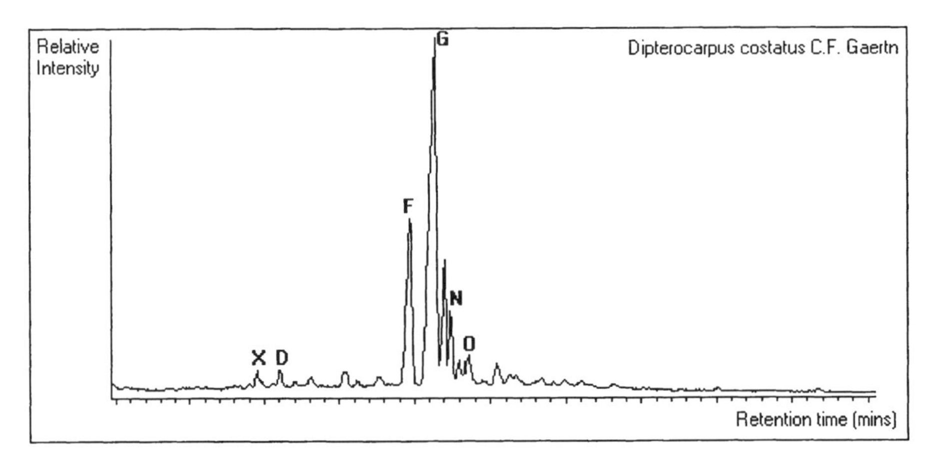


*Figure 4. Triterpenoid region of chromatogram of* Dipterocarpus costatus *resin*

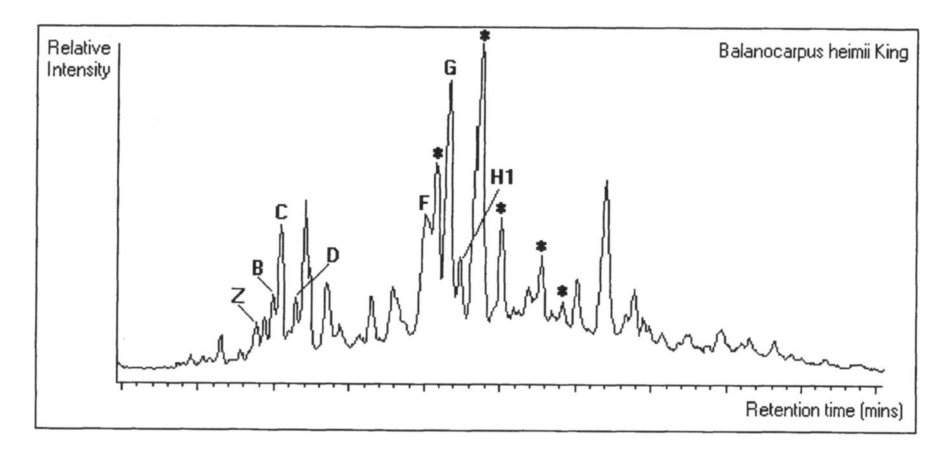


*Figure 5. Triterpenoid region of chromatogram of* Balanocarpus heimii *resin*

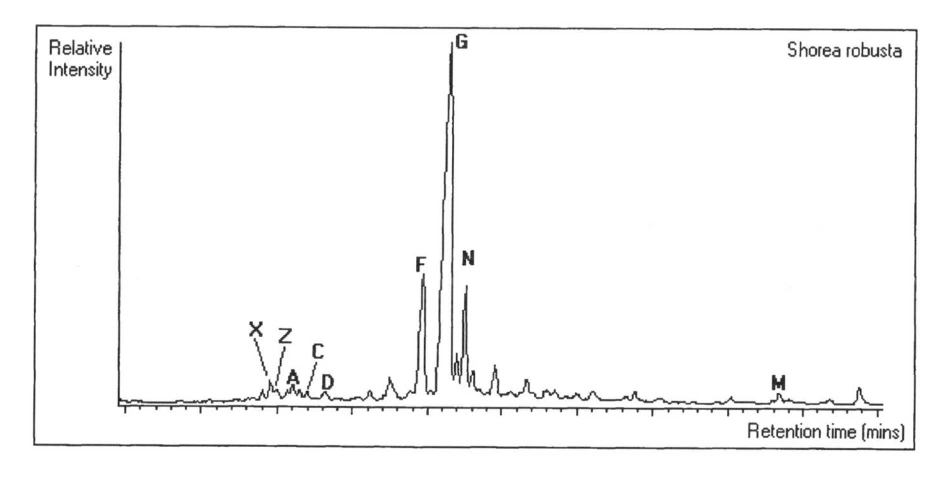


*Figure 6. Triterpenoid region of chromatogram of* Shorea robusta *resin*

Some of the identifications listed in Table II for peaks from H1 to M are tentative but many of these compounds are not well separated from one another resulting in overlapping mass spectra. All of the compounds display characteristics of the oleanane or ursane series. Several peaks, including those labelled with an asterisk (*), exhibit a 133/203/262 *m/z* pattern of fragments suggesting pentacyclic 28-methyl esters (*26*).

With the number of modern resin samples available for analysis, it has not yet been possible to differentiate between the different genera based purely on

**Table II. Identities and m/z values of triterpenoid peaks from modern Dipterocarp resins**

| *Peak* | *Compounds observed in Dipterocarp resins* | *m/z values of fragment ions under 70eV EI (relative intensity %)* |
|---|---|---|
| X | Dammaradienone | 69(100), 109(50), 189(22), 205(30), 424(12) |
| Z | Nor-β-amyrone | 204(100), 55(72), 105(54), 175(32), 189(56), 410(14) |
| A | Dammaradienol | 69(100), 93(48), 109(50), 189(32), 207(36), 426(9) |
| B | β-amyrin | 218(100), 189(20), 203(58), 426(3) |
| C | Nor-α-amyrone | 204(100), 55(54), 105(56), 175(32), 189(32), 410(9) |
| D | α-amyrin | 218(100), 189(24), 203(25), 426(6) |
| E | Unidentified (possibly a hopane derivative) | 189(100), 55(90), 95(92), 121(80), 207(68), 426(19) |
| F | 20,24-epoxy-25-hydroxy-dammaran-3-one | 143(100), 59(58), 125(16), 205(10), 399(14), 443(2) |
| G | Hydroxydammarenone | 109(100), 69(86), 95(43), 205(26), 313(8), 355(10), 424(9) |
| H1 | probably a pentacyclic 28-methyl ester | 203(100), 69(70), 109(60), 189(32), 262(50), 426(6), 470(4) |
| H2 | probably a pentacyclic 28-aldehyde | 203(100), 189(21), 232(20), 426(3), 440(2) |
| I | probably a pentacyclic 28-methyl ester | 203(100), 55(64), 133(58), 232(8), 262(16), 409(6), 438(4) |
| J1 | Oleanolic acid methyl ester | 203(100), 57(34), 133(54), 262(44), 470(2) |
| J2 | Ursolic aldehyde | 203(100), 57(26), 133(33), 232(8), 440(2) |
| K | Oleanonic acid methyl ester | 203(100), 105(20), 133(24), 189(30), 262(44), 452(4), 68(1) |
| L | Ursonic acid methyl ester | 262(100), 105(30), 133(72), 189(35), 203(78), 409(1), 452(4), 468(1) |
| M | probably a pentacyclic 28-methyl ester | 262(100), 75(87), 105(40), 133(72), 203(72), 498(3) |
| N | Dammarenediol | 109(100), 69(86), 135(28), 207(19), 357(4), 426(4) |
| O | probably a 20,24-epoxy-25-hydroxy-dammaran | 143(100), 59(28), 85(12), 107(12), 125(16), 397(2), 415(4), 475(1) |
| * | probably pentacyclic 28-methyl esters | all display 133 / 203 / 262 fragments and most include 468 |

their triterpenoid composition. This may become more feasible once the sesquiterpenoid component of each of the resins has also been considered.

Examination of two species of Dipterocarpus (*D. alatus* and *D. costatus*) has revealed clear differences between two resins from the same genus. This chemical inhomogeneity between the species may stem from the fact that the genus Dipterocarpus is a particularly large genus whose members range from dry, hill forest trees to evergreen lowland inhabitants (*5*).

## Triterpenoid resin from a non-Dipterocarp source

The *Canarium* resin, whilst still a triterpenoid resin, is quite different in composition to resins belonging to the Dipterocarpaceae family. *Canarium* belongs to the Burseraceae family and no compounds from the dammarane series are observed in its resin (Figure 7). It is dominated by two compounds, α-amyrin and β-amyrin (Table III), and although either of these may be detected in Dipterocarp resins, they are not the dominant constituents. In *Canarium* resin they are present in far higher abundance. Several other components are apparent, as a noticeable group of peaks after the α- and β-amyrin, but the mass spectra of most of these differ from those seen in Dipterocarp resins. It has not, as yet, been possible to positively confirm the identity of these peaks, although one peak (r) may be a pentacyclic 28-methyl ester and the two with major fragments at 234 m/z (s and t) may be saturated oleanane or ursane compounds (*27*).

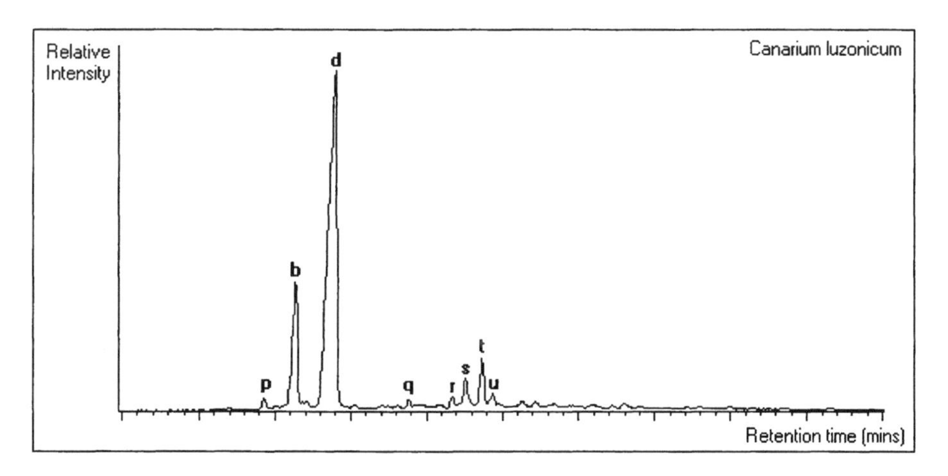


*Figure 7. Triterpenoid region of chromatogram of Canarium luzonicum resin*

**Table III. Identities and m/z values of triterpenoid peaks from modern Canarium resin**

| *Peak* | *Compounds observed in Canarium resin* | *m/z values of fragment ions under 70eV EI (relative intensity %)* |
|---|---|---|
| b | β-amyrin | 218(100), 189(20), 203(58), 426(3) |
| d | α-amyrin | 218(100), 189(24), 203(25), 426(6) |
| p | Unidentified | 218(100), 55(94), 69(78), 119(68), 133(70), 255(42), 295(16), 424(77), 485(6), 504(4) |
| q | Unidentified | 248(100), 55(90), 69(74), 95(66), 119(50), 133(52), 147(53), 207(36), 424(20), 456(19) |
| r | possibly a pentacyclic 28-methyl ester | 203(100), 189(21), 232(10), 262(23), 440(2) |
| s | possibly a saturated oleanane or ursane | 234(100), 135(40), 190(48), 273(10), 424(2), 440(2) |
| t | possibly a saturated oleanane or ursane | 234(100), 201(35), 219(34), 273(2), 424(2), 440(2) |
| u | Unidentified | 273(100), 105(48), 135(93), 175(48), 203(20), 232(62), 440(4) |

### Diterpenoid resins

The diterpenoid constituents of resins analysed are usually seen as a group of peaks eluting at around 20 minutes retention time. All of the diterpenoid resins in Table I are coniferous and come from the Pinaceae and Araucariaceae families. To date, no evidence for the use of diterpenoid resins in association with archaeological ceramics has been found.

### Archaeological samples

Samples of a visible surface residual material has been scraped from the surface of archaeological potsherds for analysis. The location and physical appearance of the resin on the sherds varied a great deal. Although material is available from several sites only two, Spirit Cave and Noen U-Loke, both in Thailand are discussed here since samples from these sites have been both analysed to determine their chemical composition and submitted for radiocarbon dating.

The resin from the Hoabinhian upland rockshelter site of Spirit Cave in Northwest Thailand tightly adhered to the underlying ceramic and appeared as a yellow/brown semi-translucent 'glassy' coating on both interior and exterior surfaces. The coating was fairly thick and seemed quite intact in places, particularly when it was preserved in indentations left by cord-marking on the sherd exterior.

In contrast, the potsherds from Noen U-Loke, an Iron Age site in the Mun River valley in Northeast Thailand apparently had an interior deposit only. This was most commonly seen as a very friable cream/white open-textured layer, sometimes overlain by a layer of soil. Some sherds bore areas of a darkened (perhaps burnt) deposit, sometimes of slightly greasy appearance.

At this stage, possible residues absorbed into the ceramic matrix of potsherds have not been investigated, although it is likely that resin may, if in contact with a pot whilst in a liquid or semi-liquid state, penetrate the ceramic to a degree. If this is the case, traces of the resin may be obtained for analysis by grinding a small section of a sherd to powder and extracting the organic material with solvent.

## Preliminary Results and Discussion

Where residues exhibited a terpenoid component, these were most commonly triterpenoid in nature, although in a few cases some sesquiterpenoids also survived. Closer examination of the mass spectra of terpenoid components of the archaeological samples showed similarities with the compounds found in modern resins of the Dipterocarpaceae family.

### Spirit Cave, Thailand

Chromatograms of samples from resin-coated pottery sherds collected at Spirit Cave all show strong similarities, whether samples are taken from the interior or exterior resin coatings. A typical example is shown below (Figure 8). Peak identities are listed in Table II. The resin coatings have a triterpenoid component eluting at around 30 minutes retention time and show clear similarities in composition to modern resins from the Dipterocarpaceae family, particularly those of *Dipterocarpus alatus* resin (Figure 3) and, to a lesser extent, *Balanocarpus heimii* resin (Figure 5).

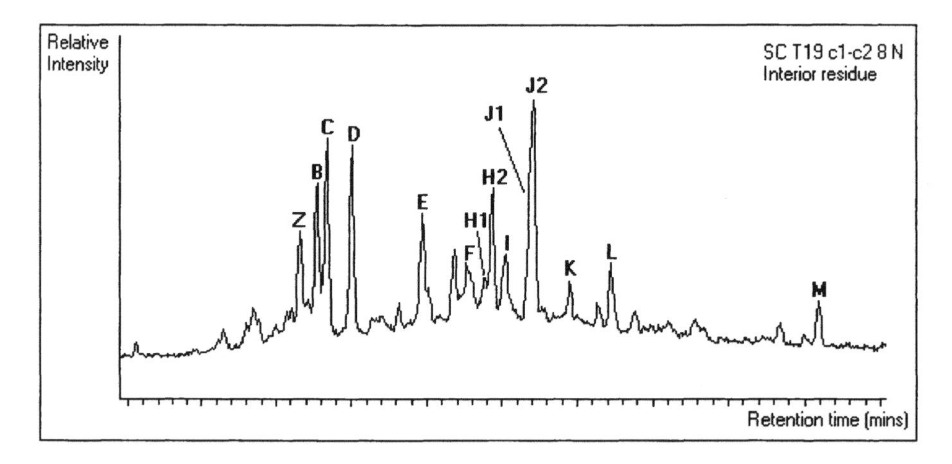


*Figure 8. Triterpenoid region of an interior resin coating from a Spirit Cave potsherd.*

Gorman (*28*), reporting the flora in the vicinity of the cave mentions two species from the Dipterocarpaceae family, both of the genus Dipterocarpus. These are *D. costatus* on the higher slopes and *D. obtusifolius* on the valley floor. However, the chromatogram from the archaeological material appears quite unlike the chromatogram of modern *Dipterocarpus costatus* (Figure 4) so it would seem unlikely that this could be the botanical source of the resin.

That neither *Dipterocarpus alatus* nor *Balanocarpus heimii* are mentioned in reports on the Spirit Cave site is unsurprising, as *D. alatus* is more commonly found at lower altitudes (up to 500m), often along rivers (*5*) and *Balanocarpus heimii* was found only in the southernmost region of Peninsular Thailand, although it may now be extinct even there (*29*). Whilst it is apparent that the archaeological resin has a triterpenoid component and closely resembles resins from the Dipterocarpaceae family in composition, at this stage it is not practicable to further narrow the botanical source down to a specific genus, far less a particular species.

## Noen U-Loke, Thailand

Not all of the residues that looked as if they might be resinous proved to be so on analysis. In particular, the friable layers seen on some sherds from Noen U-Loke which looked quite similar, even under low power microscopy, to known resin coatings from the same site yielded practically no organic matter and were probably merely an accumulation of soil or silt from the burial environment. However, where the deposits did have an organic makeup, chromatograms from samples of interior resin coatings on potsherds from Noen U-Loke again show clear similarities from one to another. As a typical example (Figure 9) shows, the archaeological residue samples exhibit a triterpenoid component. Some similarities between this material, the samples from Spirit Cave and modern Dipterocarp resins can be observed.

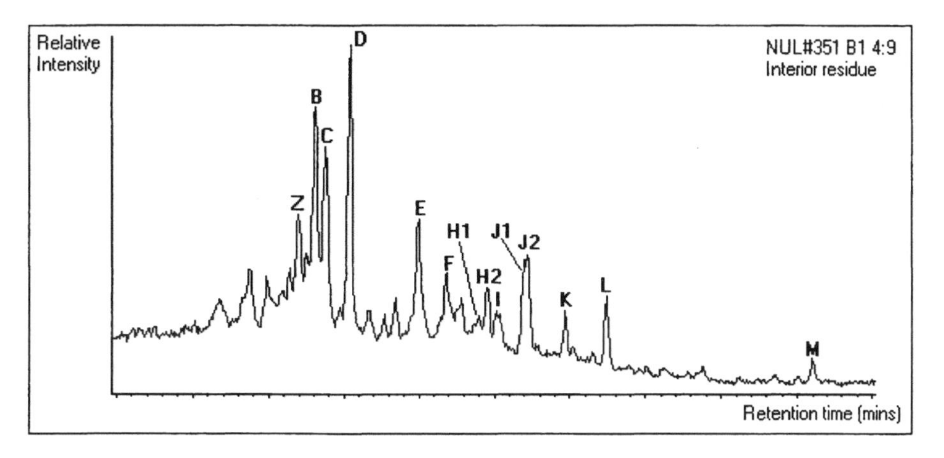


*Figure 9. Triterpenoid region of an interior resin coating from a Noen U-Loke potsherd.*

However, the samples from Noen U-Loke do not demonstrate a clear-cut correspondence to the modern Dipterocarp resins as neither hydroxydammarenone (G), which is not detected at all in these samples, nor ursolic aldehyde (J2) dominate. Instead, α-amyrin (D) is the usually largest peak by far. Although this is the case, the composition of the archaeological material is quite dissimilar to the triterpenoid Canarium resin, but does exhibit a range of the terpenoid components present in Dipterocarp resins suggesting that this family may be the botanical source of the resin used as a coating on these ceramics.

## Radiocarbon dating

Radiocarbon dating techniques focus on the point at which the material dated ceases any exchange of carbon with the biosphere; the radioactive isotope of carbon, $^{14}C$, is no longer replenished and begins to decay exponentially (*30*). In a given sample, the stable carbon isotopes, $^{12}C$ and $^{13}C$, do not decay so their concentration will remain constant but the amount of $^{14}C$ in the sample decreases as it decays over time. The abundance of $^{14}C$ can be measured either by beta emission activity or by atom counting and, based on the result of this, an 'age' can then be calculated for the material submitted for dating (*31*).

The exchange of carbon with the biosphere ceases at the point when the resin exudes from the plant, whether this is due to natural processes or to man's intervention in tapping the resin. Nicholas (*32*) reports experimental work on pine resins during the 1950s which suggests that resins remain involved in metabolic activity whilst held in resin ducts in the structure of the tree and consequently continue to be involved in the carbon cycle whilst within the plant tissues. Hence dates obtained from resins actually correspond to the date the substance exuded from the tree.

Since a resin coating is unlikely to be applied except during the functional lifetime of a pot, it can be regarded as an integral part of the pot. This gives us full certainty of association between the material to be dated, the resin, and the object of interest, the pottery, and would suggest that dating the resinous coatings on potsherds may offer a direct method of dating the ceramics themselves.

Having first determined that the visible residues on potsherds from Noen U-Loke and Spirit Cave were organic, i.e. carbon based, and having identified the deposits as resinous, small samples of the material were taken for accelerator mass spectrometry (AMS) radiocarbon dating. A major advantage of the AMS radiocarbon dating technique over conventional beta counting is that a considerably smaller mass of sample is required. For this type of material the amount required is typically about 20–50 mg of visible deposit. The optimum amount of carbon following pretreatment is approximately 2-4 mg (*33*).

The application of a resin coating on pottery often results in quite a thin layer and, particularly where sherds are small and coated on one surface only, it can prove problematic to collect enough material. However, sufficient material was present on at least two sherds to enable more than one sample to be collected. This allowed the repeatibility of dates using this material to be assessed for both Noen U-Loke and Spirit Cave.

# Methods

Small samples of the visible organic residues were collected using a scalpel to dislodge the resin from the surface of the potsherds. The removal of this material from the sherd surface was carried out under under low power microscopy to minimise obvious contamination from sources of modern carbon (e.g. rootlets etc. from the burial environment, modern adhesives, traces of labeling varnish, small fibres from packing materials, etc.). All samples were submitted for dating in glass vials with foil-lined caps. Analysis using GC-MS had highlighted a small amount of organic contamination due to modern plasticisers and the dating laboratory was duly notified of this. Pretreatment of the samples at the radiocarbon laboratory involved solvent extraction of the material to specify the source of the carbon. The chloroform soluble fraction was retained since it contains the resin. Targets were prepared using this chloroform extract and these were submitted for dating (*34*).

### Archaeological samples

Three samples were submitted for radiocarbon dating from Noen U-Loke. Each sample comprised a small amount of the visible resin coatings on the interior of potsherds. The material from this site was chosen to assess the validity of radiocarbon dates acquired using plant resin samples by comparing them with a published series of well-stratified dates from the site. Of these samples, two came from a single potsherd and acted as a check that the dates were repeatable and that resin coatings are relatively homogeneous.

Two further samples were submitted from the rockshelter site of Spirit Cave, again obtained from a single sherd. One of the samples collected was from an exterior resin coating, the other from an interior resin coating. These samples were chosen to try to resolve uncertainties concerning an early date originally suggested for the pottery and, coming from one sherd, were used as a further check on the repeatability of dates using resin samples

# Radiocarbon results

The radiocarbon dates obtained from samples of resinous coatings on potsherds from the two archaeological sites in Southeast Asia, Noen U-Loke and Spirit Cave, are listed in Table IV.

**Table IV. AMS radiocarbon dates from samples of resin coatings on pottery.**

| *Oxford Ref* | *Site* | *Sample reference* | *Date BP (uncal)* | *Calibrated dates* |
|---|---|---|---|---|
| OxA-10268 | Noen U-Loke | NUL#208 A2 4:5 4 | 1900 +/- 37 | 20AD (95.4%) 230AD |
| OxA-10269 | Noen U-Loke | NUL#208 A2 4:5 4 | 1861 +/- 35 | 70AD (95.4%) 250AD |
| OxA-10270 | Noen U-Loke | NUL#351 B1 4:9 | 2149 +/- 35 | 20BC (95.4%) 230BC |
| OxA-10271 | Spirit Cave | SC T19 c1-c2, 8 (N) | 3042 +/- 37 | 1410BC (90.9%) 1210BC<br>1200BC (1.7%) 1190BC<br>1180BC (1.1%) 1160BC<br>1140BC (1.6%) 1130BC |
| OxA-10272 | Spirit Cave | SC T19 c1-c2, 8 (N) | 2995 +/- 40 | 1390BC (94.2%) 1110BC<br>1100BC (1.2%) 1080BC |

Isotopic fractionation was measured by the radiocarbon laboratory and corrected for. Dates were reported both uncalibrated, in years BP, and calibrated, using the 'OxCal' computer program (version 3.5) (*35*). The atmospheric data used in calibration was from 'INTCAL 98'(*36*)

## Discussion

All three dates from Noen U-Loke correlate well with the existing series of dates from the site (*37, 38*) (Figure 10). This suggests that dates from resin samples are valid and reliable. The dates obtained from two separate samples taken from a single sherd are in good agreement with each other, indicating that the dates are repeatable and further supporting evidence from the analysis that suggests that the resin coatings are reasonably homogeneous.

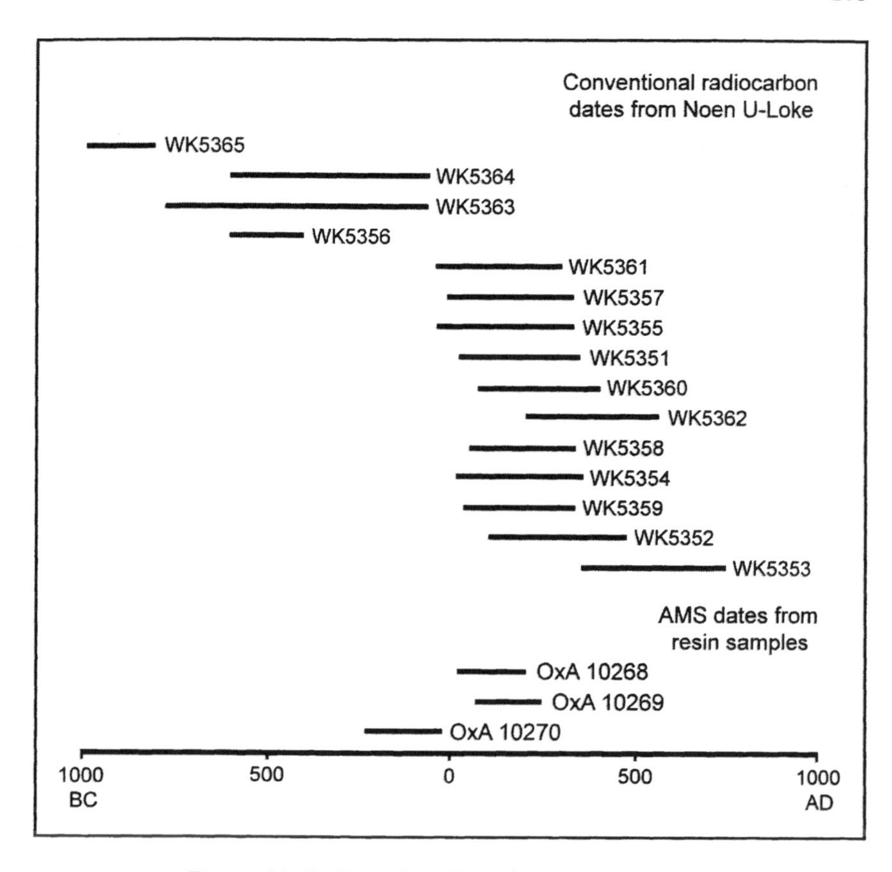


*Figure 10. Radiocarbon dates from Noen U-Loke*

However, the two dates from Spirit Cave are not in agreement with a published series of bamboo-charcoal dates (Figure 11) from the upper levels of the site, which suggested that the ceramics (recovered from Layer 1 and the surface of Layer 2, Cultural Level II) were of late Hoabinhian origin (*28, 39*)

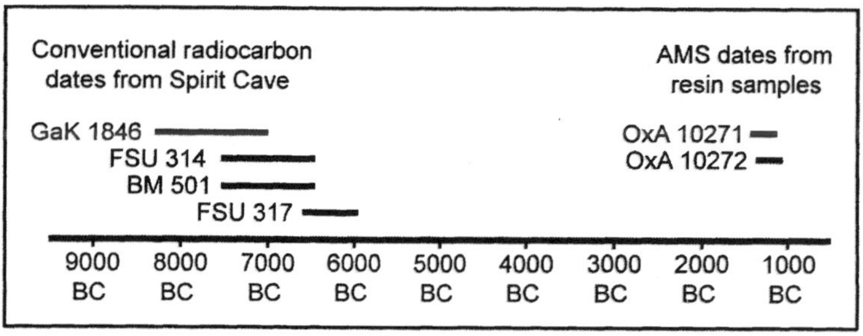

Conventional radiocarbon dates from Spirit Cave
AMS dates from resin samples
GaK 1846
FSU 314
BM 501
FSU 317
OxA 10271
OxA 10272
9000 BC
8000 BC
7000 BC
6000 BC
5000 BC
4000 BC
3000 BC
2000 BC
1000 BC

*Figure 11. Radiocarbon dates from Spirit Cave*

Neither does it appear that the potsherds were introduced to the site by disturbance of later log coffins (thought to be Iron Age), but may suggest a continuation in the use, possibly intermittent, of the rockshelter during the Late Neolithic-Bronze Age. However, this particular sherd comes from the surface of the site rather than within a recognised horizon and may post-date sherds found deeper in the sequence. The two dates (from a single sherd) are in good agreement with one another, supporting the view that dates from resins are repeatable and that both coatings were applied to the ceramic contemporaneously.

Ostensibly, the radiocarbon dates obtained from samples of resin appear to be reliable and repeatable and, since a resin coating is unlikely to be applied except during the functional lifetime of a pot, would seem to suggest that resinous coatings on potsherds may offer a direct method of dating the ceramics. Whilst these initial results are promising, further dates from resin-coated ceramics are necessary before the results can be fully supported or refuted and plans are underway for additional samples to be dated.

On a cautionary note, it would be prudent to identify the botanical source of any Southeast Asian resin samples before submitting them for radiocarbon dating. Whilst it might be assumed that resin collected for use on ceramics was freshly exuded from the tree and hence roughly contemporary with the pot to which it is applied, there is a history of certain types of resin, particularly *Agathis* resin (*40*), being collected in bulk from the ground beneath old trees or from areas where large trees once grew. Since radiocarbon techniques date the cessation of exchange of carbon with the biosphere, a ground collected resin, potentially exuded many years earlier, might considerably predate the ceramic object to which it was applied. Radiocarbon dates obtained from deposits of these ground-collected resins used as coatings, adhesives etc. may mislead as they would suggest a greater age for an object than was actually the case. Whilst this appears to have been a more widespread practice in the case of resins such as *Agathis*, Burkill (*41*) does make mention of dammar resins from some members of the Dipterocarpaceae family being collected “in the shape of accumulations near dead tree stumps and dug up from the soil where no trace of the parent tree remain”. Hence care should be taken in interpreting radiocarbon dates obtained from any resins known to have been collected in this manner at any time in the past.

**Acknowledgements**

We express our gratitude to the following people and organisations. The Natural Environment Research Council, UK for the award of a research studentship and a supplementary award for radiocarbon analysis, which supports the work being carried out. The Research Laboratory for Archaeology and the History of Art, Oxford, UK for undertaking AMS dating of resinous samples. The Royal Botanic Gardens, Kew, UK for supplying modern reference samples and access to their research libraries. The Francis Raymond Hudson Memorial Fund for their kind assistance towards conference travel expenses. Prof. Charles Higham, University of Otago, New Zealand for access to material from excavations at Noen U-Loke. Dr David Bulbeck, Australian National University, Australia for access to material from excavations at Bola Merajae. Carl Lander and Yvonne Forte for work previously carried out at Bradford on some archaeological material included in this study. Dr Joyce White, of MASCA, University of Pennsylvania, USA for her interpretation of background material.

## References

1. Gianno, R., "Semelai Culture and Resin Technology"; Memoir 22; Connecticut Academy of Arts and Sciences: New Haven, CT. 1990.
2. Schiffer, M.B. In *Archaeometry '88: Proceedings of the 26th International Archaeometry Symposium*; Farquhar, R.M.; Hancock, R.G.V; Pavlish, L.A., Eds.; Archaeometry Laboratory, University of Toronto: Toronto, 1988; pp 23-29.
3. Schiffer, M.B. *J. Arch. Sci.*, **1990**, 17, 373-381.
4. Whitmore, T.C. *Tropical Rain Forests of the Far East*, 2nd ed.; Clarendon Press: Oxford, 1984.
5. Smitinand, T.; Santisuk, T.; Phengklai, C. *Thai Forestry Bulletin.* **1980**, 12, 1-133.
6. Vidal, J.E. In *Tropical Botany*; Larsen, K.; Holm-Nielsen, L.B., Eds.; Academic Press: London, 1979, pp 109-123.
7. Gianno, R.; Erhardt, W.D.; Von Endt, D.W.; Hopwood, W.; Baker, M.T. In *Organic Contents of Ancient Vessels: Materials Analysis and Archaeological Investigation, MASCA Research Papers in Science and Archaeology;* Biers, W.R.; McGovern, P.E., Eds.; MASCA: Philadelphia, 1990; Vol. 7; pp 59-67.
8. Gianno, R.; Von Endt, D.W.; Erhardt, W.D.; Kochummen, K.M.; Hopwood, W. In *Recent Advances in the Conservation and Analysis of*

*Artefacts*; Black, J., Ed.; Institute of Archaeology: London, 1987; pp 229-238.
9. Skibo, J.M. *Pottery Function: A Use-Alteration Perspective*; Plenum Press: New York, 1992.
10. Ellen, R.F.; Glover, I.C. *Man* **1974**, *9*, pp 353-379.
11. May P.; Tuckson M. *The Traditional Pottery of Papua New Guinea*; Bay Books: Kensington, New South Wales, 1982.
12. Harborne, J.B. In *Ecological Chemistry and Biochemistry of Plant Terpenoids; Proceedings of the Phytochemical Society of Europe 31*; Harborne, J.B.; Tomas-Barberan, F.A., Eds.; Clarendon Press: Oxford. 1991; pp 399-426.
13. Hanson, J.R. In *Chemistry of Terpenes and Terpenoids*; Newman, A.A., Ed.; Academic Press Inc. (London) Ltd.: London, 1972; pp 1-10.
14. Loomis, W.D; Croteau, R., In *The Biochemistry of Plants: A Comprehensive Treatise, Vol. 4. Lipids: Structure and Function*; Stumpf, P.K., Ed.; Academic Press: New York, 1980; pp 363-418.
15. Herbert, R.B. *The Biosynthesis of Secondary Metabolites*, 2nd ed; Chapman and Hall: London, 1989.
16. Torssell, K.B.G., *Natural Product Chemistry: A Mechanistic, Biosynthetic and Ecological Approach*, 2nd ed; Swedish Pharmaceutical Press: Stockholm, 1997.
17. Mills, J.S.; White, R. *The Organic Chemistry of Museum Objects*; Butterworth-Heinemann Ltd: Oxford, 1994.
18. Thomas, B.R. In *Phytochemical Phylogeny*; Harborne, J.B., Ed.; Academic Press: London, 1970; pp 59-79.
19. Bissett, N.G.; Diaz-Parra, M.A.; Ehret, C.; Ourisson,G. *Phytochem.*, **1967**, *16*, 1395-1405.
20. Cotterrell, G.P.; Halsall, T.G.; Wigglesworth, M.J. *J. Chem. Soc.*, **1970**, 739-743.
21. Thomas, B.R. In *Organic Geochemistry: Methods and Results*; Eglinton, G. ; Murphy, M.T.J., Eds.; Springer-Verlag: Berlin, 1969; pp 599-618.
22. Ames, G.R.; Chandra Rajan, M.; Matthews, W.S.A. *Trop. Sci.*, **1969**, *XI (3)*, 170-176.
23. Green, C.L.; Keeble, B.; Burley, J. *Trop. Sci.*, **1975**, *17 (3)*, 165-174.
24. Serpico, M. In *Ancient Egyptian Materials and Technology*; Nicholson, P.T.; Shaw, I., Eds.; Cambridge University Press: Cambridge, 2000; pp 430-474.
25. van der Doelen, G.A. PhD Thesis, FOM Institute for Atomic and Molecular Physics, Amsterdam, 1999.
26. de la Rie, E.R. PhD Thesis, University of Amsterdam, 1988.

27. Budzikiewicz, H.; Wilson, J.M.; Djerassi, C. *J. Am. Chem. Soc.*, **1963**, *85*, 3688-3699.
28. Gorman, C. *Asian Perspectives*, **1972**, *13*, 79-107.
29. Soerianegara, I.; Lemmens, R.H.M.J. *Timber trees: Major commercial timbers*; Plant Resources of South-East Asia No. 5(1), Pudoc Scientific Publishers: Wageningen, 1993.
30. Bowman, S. *Radiocarbon Dating (Interpreting the Past)*; British Museum Publications Ltd.: London, 1990.
31. Aitken, M.J. *Science-based Dating in Archaeology*; Addison Wesley Longman Ltd.: Harlow, 1990.
32. Nicholas, H.J. In *Biogenesis of Natural Compounds*: Bernfeld, P., Ed.; Pergamon Press: Oxford, 1967; pp 829-901.
33. Hedges, R.E.M.; Law, I.A.; Bronk, C.R.; Housley, R.A., *Archaeometry*, **1989**, 31(2), pp 99-113.
34. Higham, T.F.G. 2001, Personal communication.
35. Bronk Ramsey, C. *Radiocarbon*, **1995**, *37(2)*, 425-430.
36. Stuiver, M.; Reimer, P.J.; Bard, E.; Beck, J.W.; Burr, G.S.; Hughen, K.A.; Kromer, B.; McCormac, G.; van der Plicht, J.; Spurk, M. *Radiocarbon*; **1998**, *40(3)*, pp 1041-1083.
37. Higham, C.F.W. ; Thosarat, R. *Prehistoric Thailand: from early settlement to Sukhothai*; Thames Hudson Ltd.: London, 1998; pp 27-29.
38. Higham, C.F.W.; Thosarat, R. *Proceedings of the 16th IPPA Congress, Melaka*; **1998**, in press.
39. Erlich, R.W. In *Chronologies in Old World Archaeology*; University of Chicago Press: Chicago, 1992; pp 475-515.
40. Howes, F.N. *Vegetable Gums and Resins*; Chronica Botanica Company: Waltham, MA. 1949.
41. Burkill, I.H. *A Dictionary of the Economic Products of the Malay Peninsula;* Crown Agents for the Colonies: London, 1935; Vol I. (A-H).

## Chapter 8

# Analysis of Glass Beads and Glass Recovered from an Early 17th-Century Glassmaking House in Amsterdam

**K. Karklins[1], R. G. V. Hancock[2], J. Baart[3], M. L. Sempowski[4], J.-F. Moreau[5], D. Barham[6], S. Aufreiter[6], and I. Kenyon[7]**

[1]Parks Canada, Ontario Service Centre, Ottawa, ON K1A 0M5, Canada
[2]SLOWPOKE-2 Facility, Department of Chemistry and Chemical Engineering, Royal Military College, Kingston, ON K7K 7B4, Canada
[3]Archaeologie, Stedelijk Beheer, Amsterdam, Netherlands
[4]Rochester Museum and Science Center, Rochester, NY 14603
[5]Laboratoire d'Archéologie et Département des Sciences Humaines, Université du Québec à Chicoutimi, Chicoutimi, PQ G7H 2B1, Canada
[6]Department of Chemical Engineering and Applied Chemistry, University of Toronto, Toronto, ON M5S 3E5, Canada
[7]Ontario Heritage Foundation, Toronto, ON M5C 1J3, Canada

A variegated sample of two hundred and ninety glass beads and glass fragments recovered from the wasters of a glassmaking house in Amsterdam have been analysed by instrumental neutron activation analysis to determine their bulk elemental compositions. Since the archaeological context reveals that the material was deposited between 1601 and 1610, the recovered data provide a chronological anchor for beads of the same chemistry on archaeological sites in North America. Hence, knowledge of these Dutch bead chemistries is significant not only in determining the source of glass trade beads found in North America, and perhaps elsewhere in the world, but also in establishing the temporal locus of early archaeological sites with limited European artifacts save glass beads.

Since glass trade beads have been recovered from many aboriginal, contact and post-contact sites in northeastern North America, there has been a long-term interest amongst archaeologists to classify, date, and interpret them. The earliest classification system was based on form (*1*) but this was not found to be practicable by later researchers. Consequently, in 1970, Kidd and Kidd (*2*) published their taxonomic system for glass beads which segregated specimens on the basis of manufacture, followed by shape, diaphaneity, colour, and size. Their system was later significantly expanded by Karklins (*3*).

The first reported chemical analyses of glass beads (*4*), using energy dispersive x-ray fluorescence analysis (EDXRF), showed that drawn glass beads were made from soda-rich glasses while wound beads were made from potash-rich glasses. Subsequent studies, using non-destructive instrumental neutron activation analysis (INAA), have been conducted on blue glass beads from northeastern North America (*5, 6, 7, 8, 9*), and white glass trade beads from northeastern North America (*10, 11, 12*). These studies have shown that the major and trace element chemistries of glass beads may be used to sort similar looking beads into chemical groups, using their elemental concentration fingerprints. These elemental fingerprints have proven to been useful in both distribution studies and in chronological studies of glass trade beads.

In many cases, from historic records (e.g. *13*), it has been possible to link non-specific traders (French, Dutch, and English) with beads found in different geographical regions. Although we tend to assume the opposite, there is no direct link between the nationality of the trader and the source country, or city, of the articles traded. The determination of the nation, city, or, most optimistically, specific glass beadmaking house(s) that made the beads traded into North America is the next major step in elucidating the nature of a small part the Europe-North America trading system. So, a small step in this direction comes from the analysis of glass beads and glass fragments excavated from the wasters of a specific glassmaking house in Amsterdam that was in operation from 1601 to 1610. This type of study allows us to generate both the chronological beginning and the potential production source of specific chemistries of glass beads that were traded to North America, and indeed, starting in the 1620s, around the world.

## Site

Rescue excavations at two contiguous sites (Asd/Kg9 and Kg10) on the Keizersgracht canal between Wolvenstraat and Hartenstraat in downtown Amsterdam uncovered extensive glass- and beadmaking waster deposits (*14*). The larger of the two sites (Asd/Kg10) was uncovered in 1981 by the municipal Department of Public Works while installing a sewer line along the east side of the canal. Glassmaking debris was found to cover a region at least seventy metres long. Recovered by carefully washing and sorting the fill, the beads and beadmaking refuse were found to be concentrated in an area about two metres across and at a depth of four metres below street level. Over 50,000 specimens were recovered, including finished beads, malformed rejects, and tube fragments, all of drawn manufacture. The deposit also contained chunks of vari-coloured glass, numerous coloured glass rods, crucible remnants, and a variety of drinking-glass fragments, all indicating that this was material from a local glasshouse. The archaeological context, as well as the associated ceramics and glassware, reveals that the glasshouse operated between 1601 and 1610.

## Experimental Methods

Two hundred and ninety drawn glass beads and fragments of varying colour, structure (flashed, cored and uncored), sizes, and shapes from site Asd/Kg10 were analyzed non-destructively using instrumental neutron activation analysis at the SLOWPOKE Reactor Facility of the University of Toronto. These beads had to be neutron-irradiated as little as possible in order to minimize the build-up of radioactivity from $^{122}Sb$ (half-life 2.75 days) and from $^{124}Sb$ (half-life 60.9 days) in the Sb-rich beads, so that they could be returned to their owners within a reasonable amount of time.

Beads of mass 5-10 mg were first cleaned ultrasonically, as required. They were stored individually in 1.2 mL polyethylene vials, were irradiated serially for five minutes at a neutron flux of $1.0 * 10^{12}$ neutrons.cm$^{-2}$.sec$^{-1}$. Five to seven minutes after irradiation, the induced radioactivity was counted for five minutes using a hyper-pure germanium detector-based gamma-ray spectrometer. This produced analytical concentration data for cobalt (Co), tin (Sn), copper (Cu), sodium (Na), aluminium (Al), manganese (Mn), chlorine (Cl) and calcium (Ca). The samples were recounted for five to thirty-three minutes the next day to measure the concentrations of the longer-lived radioisotopes of Na, arsenic (As), antimony (Sb) and potassium (K). The Na measurements were used to link both counts. Elemental concentrations were calculated using the comparator method, based on elemental standards. Beads of larger masses were irradiated at suitably lower neutron fluxes to make just enough radioactivity for reasonable chemical analyses.

## Results and Discussion

The colours of the analyzed glasses were turquoise blue (14 samples), royal blue (12 samples), white (45 samples), red (135 samples), black (52 samples), colourless (21 samples) and gold (11 samples). Table I lists the samples by colour, shape, and form. The gold coloured beads were made from leaded glasses. All of the other samples were soda-lime-silica glasses, with variable amounts of potash. The complete data set is available from the authors.

### Turquoise Blue Glasses

Twelve round turquoise blue beads (Table II; samples 1-12) are characterised by high Na (11.5 % to 13.0 %) and high Cl (1.49 to 2.01 ), and low K (1.7 % to 2.8 %), low Ca (3.3 % to 5.4 %), and low Al (0.48 % to 0.64 %). In terms of the gross glass chemistry, these beads are chemically similar to Ontario Bead Period II (≈1600-1620) and Bead Period III (≈1620-1650) turquoise blue beads, as reported in Hancock *et al.* 1994.

The two scrap samples (283 and 288) are Sn-rich glasses of different basic chemistries. Scrap 283 has a curious K-rich/Na-rich composition, which is close to the chemistry of colourless scrap 263 (see Table III) and which may perhaps be related to enamel work.

### Royal Blue Glasses

Twelve royal blue (the Kidds' "brite navy") beads (Table II, samples 73-84) were all lime-rich (5.6 to 7.6 % Ca) soda glasses, coloured with Co, with about 2-3 times as much As, as was observed by Hancock *et al.* 2000. Relative to the turquoise blue beads, these cobalt blue beads are lower in Na and Cl and higher in Ca, Al, and much higher in Mn. Either the extra Mn was added intentionally or it came into the glass with the Co-colouring material. Ten of the twelve beads (samples 73-82) are chemically consistent, forming a single chemical group, while beads 83 and 84 are each quite different.

**Table I. Summary Data**

| Colour | # | Description | Kidd & Kidd ID |
|---|---|---|---|
| turquoise | 1-12 | Round | IIa1 |
| blue-green | 283, 288 | Fragments | |
| royal blue | 73-84 | Round | IIa55 |
| white | 13-24 | clear light grey/op white/cl light grey | IVa11 |
| | 25 | opaque white fragment | |
| | 193-196 | tubular fragments | |
| | 245-256 | thin red and blue striped tube frags | Ib7 |
| | 258, 260-263, 268 | flat sheet fragments | |
| | 275-277 | bead fragments | |
| | 281-290 | fragments (less 283, 287, 288) | |
| redwood | 26-37 | round/circular | IIa1/IIa2 |
| | 51-62 | grey/green cored circular (cc) | IVa2 |
| | 85-101 | red solid tubes (tu) | Ia1(a) |
| | 102-114 | black cored tubes | IIIa1 |
| | 115-127 | tubes flashed in clear glass | Ia1(b) |
| | 133-151 | cored barrels/circulars | IIa1/IIa2 |
| | 152 | cored circular | IIa2 |
| | 153-172 | cored round/circular | IVa2/IVa3 |
| | 173-192 | cored circular/round | IVa5/IVa6 |
| | 237-244 | broad blue + white stripes | |
| black | 39-50 | Round | IIa6 |
| | 197-216 | round black/grey | IIa6 |
| | 217-236 | circular black/grey | IIa7 |
| colourless | 128-132 | Wasters | |
| | 257-270 | cullet - flat sheets | |
| | 271-280 | lumps and chunks | |
| | 287 | Fragment | |
| gold | 38 | opaque light | IIa17 |
| | 63-72 | light tubes | Ia7 |

**Table II. Data for the Blue Glasses**

| # | Al % | Ca % | Cl % | Co ppm | Cu % | Mn ppm | K % | Na % | Sn ppm | As ppm | Sb ppm |
|---|---|---|---|---|---|---|---|---|---|---|---|
| **Round turquoise blue beads** | | | | | | | | | | | |
| 05 | 0.41 | 3.4 | 2.01 | <17 | 0.79 | 126 | 2.9 | 12.0 | <650 | <58 | <62 |
| 01 | 0.48 | 3.4 | 1.93 | <54 | 1.04 | 225 | 1.7 | 12.1 | <850 | <72 | <110 |
| 06 | 0.54 | 4.4 | 1.49 | 78 | 0.92 | 238 | 1.7 | 11.8 | <790 | <74 | <100 |
| 11 | 0.53 | 5.1 | 1.74 | 79 | 0.98 | 255 | 2.1 | 12.6 | <760 | <75 | <100 |
| 12 | 0.62 | 5.4 | 1.79 | 70 | 0.97 | 291 | 1.8 | 12.7 | <820 | 200 | <91 |
| 08 | 0.53 | 3.4 | 1.76 | 40 | 1.15 | 317 | 1.8 | 11.8 | 2960 | <67 | <86 |
| 10 | 0.62 | 3.7 | 1.57 | <60 | 1.44 | 447 | 1.7 | 11.5 | <1000 | <70 | <110 |
| 02 | 0.64 | 3.8 | 1.61 | 62 | 0.79 | 473 | 2.0 | 12.3 | 1970 | <80 | <110 |
| 03 | 0.56 | 3.7 | 1.96 | <79 | 0.95 | 471 | 2.8 | 12.8 | <1300 | <62 | <96 |
| 04 | 0.53 | 3.9 | 1.68 | <60 | 0.88 | 580 | 2.6 | 12.6 | <1000 | <53 | <77 |
| 07 | 0.53 | 3.3 | 1.69 | 66 | 0.74 | 612 | 2.2 | 13.0 | <610 | 250 | <63 |
| 09 | 0.61 | 3.6 | 1.64 | <53 | 0.83 | 983 | 1.8 | 12.9 | <950 | 170 | <120 |
| **Turquoise blue scraps** | | | | | | | | | | | |
| 283 | 0.65 | 4.3 | 1.28 | 119 | 1.00 | 622 | 10.9 | 8.4 | 7800 | 420 | <75 |
| 288 | 0.78 | 6.9 | 0.88 | <31 | 1.03 | 5303 | 3.3 | 10.3 | 9300 | <44 | <50 |
| **Round royal blue beads** | | | | | | | | | | | |
| 73 | 0.85 | 7.6 | 0.91 | 1210 | <0.05 | 3390 | 1.9 | 10.3 | <1500 | 2610 | <100 |
| 74 | 0.85 | 7.0 | 0.80 | 1150 | <0.05 | 3370 | 1.7 | 9.9 | <1000 | 2470 | <120 |
| 75 | 0.88 | 6.8 | 0.86 | 1210 | <0.05 | 3440 | 1.7 | 9.8 | <1600 | 2460 | <110 |
| 76 | 0.87 | 7.0 | 0.86 | 1200 | <0.06 | 3180 | 2.0 | 9.9 | <1100 | 2510 | <130 |
| 77 | 0.92 | 6.9 | 0.86 | 1060 | <0.05 | 3540 | 1.7 | 9.9 | <1700 | 2340 | <110 |
| 78 | 0.88 | 7.6 | 0.80 | 1260 | <0.05 | 3480 | 1.8 | 9.7 | <980 | 2520 | <110 |
| 79 | 0.84 | 7.7 | 0.79 | 1190 | <0.05 | 3270 | 2.1 | 9.7 | <1200 | 2550 | <61 |
| 80 | 0.83 | 7.3 | 0.86 | 1100 | <0.04 | 3300 | 1.7 | 9.6 | <1100 | 2550 | <48 |
| 81 | 0.87 | 7.1 | 0.84 | 1200 | <0.04 | 3360 | 1.9 | 9.4 | <800 | 2570 | 114 |
| 82 | 0.85 | 7.0 | 0.80 | 1070 | <0.05 | 3270 | 2.0 | 9.6 | <1100 | 2570 | <58 |
| 83 | 0.97 | 7.7 | 0.78 | 610 | <0.06 | 9660 | 1.7 | 8.7 | <1300 | 3930 | 120 |
| 84 | 1.14 | 5.6 | 0.85 | 480 | <0.04 | 1320 | 2.5 | 8.5 | 2000 | 520 | <46 |

## White Glasses

Six subsets of white glass were analysed (see Table III). Samples 13-24 are "transparent light grey - opaque white - transparent light grey" circular beads (IVa11); sample 25 is an opaque white, round bead; samples 193-196 are white

Table III. White Glass Data

| # | Al % | Ca % | Cl % | Co ppm | Cu % | Mn ppm | K % | Na % | Sn ppm | As ppm | Sb ppm |
|---|---|---|---|---|---|---|---|---|---|---|---|
| **Circular white beads** | | | | | | | | | | | |
| 25 | 0.56 | 7.0 | 0.61 | <16 | <0.02 | 1850 | 3.6 | 7.4 | 68700 | <54 | <45 |
| 17 | 0.98 | 5.3 | 0.90 | <69 | <0.05 | 2770 | 3.7 | 9.1 | 53800 | <38 | <36 |
| 19 | 0.97 | 6.3 | 0.91 | <25 | <0.04 | 3220 | 3.2 | 8.7 | 46400 | <53 | <53 |
| 18 | 0.98 | 6.0 | 1.02 | <32 | <0.06 | 4930 | 4.1 | 9.3 | 14200 | <39 | <37 |
| 22 | 0.97 | 6.7 | 0.97 | <31 | <0.05 | 5430 | 3.8 | 9.4 | 15300 | <78 | <63 |
| **White bead fragments** | | | | | | | | | | | |
| 193 | 0.96 | 5.8 | 0.88 | <21 | <0.04 | 2580 | 4.7 | 9.0 | 51800 | <62 | <58 |
| 282 | 1.02 | 5.9 | 0.66 | 62 | <0.06 | 2820 | 3.6 | 8.1 | 65300 | 260 | 114 |
| 195 | 1.11 | 6.2 | 0.94 | <60 | <0.06 | 3170 | 4.2 | 10.2 | 44000 | <69 | <66 |
| 196 | 0.81 | 5.2 | 0.88 | 240 | <0.06 | 3640 | 4.9 | 9.3 | 19900 | 280 | <52 |
| 285 | 0.94 | 7.2 | 0.75 | <34 | 0.17 | 6510 | 4.1 | 9.9 | 5400 | <41 | <43 |
| **Cullet scraps – flat sheets** | | | | | | | | | | | |
| 260 | 0.92 | 5.8 | 0.80 | 65 | 0.18 | 1490 | 3.0 | 9.0 | 36300 | 287 | 172 |
| 261 | 1.01 | 6.3 | 0.94 | <80 | <0.04 | 2250 | 3.6 | 10.3 | 16700 | 182 | <70 |
| 263 | 0.48 | 4.5 | 0.64 | 150 | 0.21 | 3440 | 7.8 | 8.4 | 26100 | <72 | <74 |
| 258 | 1.35 | 6.6 | 0.83 | <94 | <0.06 | 3690 | 3.5 | 11.0 | 4400 | 210 | <61 |
| **Bead fragments – thin red and blue stripes on white body** | | | | | | | | | | | |
| 247 | 0.85 | 5.8 | 0.71 | <56 | 0.15 | 2630 | 4.6 | 8.4 | 33700 | 220 | <120 |
| 246 | 0.96 | 5.6 | 0.83 | <78 | <0.06 | 3590 | 4.4 | 9.0 | 28100 | 160 | <59 |
| 248 | 0.86 | 5.7 | 0.73 | <84 | <0.06 | 4560 | 4.1 | 8.2 | 37000 | 210 | <51 |
| 249 | 0.90 | 5.5 | 0.86 | 110 | 0.15 | 5380 | 3.5 | 9.4 | 39800 | 210 | <59 |
| **Wasters** | | | | | | | | | | | |
| 276 | 0.99 | 6.4 | 0.91 | <25 | <0.05 | 2780 | 3.9 | 8.7 | 48300 | <39 | <44 |
| 277 | 0.39 | 6.6 | 0.92 | 99 | 0.11 | 4010 | 2.8 | 10.9 | 15400 | <68 | <79 |

tubular fragments; samples 245-256 are white, tube fragments with thin red and blue stripes (Ib7); samples 275-277, 281, 282, 284-286, 289, and 290 are white bead fragments; and samples 258, 260-263, and 268 are flat appliqué fragments. All of the white beads were opacified with Sn, at levels ranging from as low as 0.5 %, but most commonly between 1.4 % and 8.1 %.

In samples 245-256, with narrow red and dark blue stripes, the Cu (red) and Co (blue) are rarely measurable, although the As, which is associated with Co, is often measurable at low levels. This shows that for this type of multi-coloured bead, bulk elemental analysis must be used in conjunction with visual inspection of beads and bead fragments.

**Red Glasses**

The one hundred and thirty-five red glass beads were identified initially by shape. There were circular uncored (cu) beads, round uncored (ru) beads, barrel-shaped cored (bac) beads, circular cored (cc) beads, circular black cored (cbc) beads, round cored (rc) beads, round black cored (rbc) beads, tubular flashed (tf) beads, tubular uncored (tu) beads, and tubular cored black (tcb) beads.

The circular and round beads were combined into cored and uncored groups. The tubular beads were sorted into black-cored, uncored, and flashed beads. Representative data of this bead selection, together with data from fragments of beads with broad dark blue stripes on white stripes on red bodies (b/w/r) are presented in Table IV. Within each visually-defined group, the samples are sorted arbitrarily (from top to bottom) by their Mn contents. This shows that there are a number of similar, but different, chemistries within each visually identified, coarse group. With this bead group heterogeneity in mind, group means and standard deviations (Table V) must be interpreted with care.

The blue-on-white-striped red-bodied beads (b/w/r) are the most chemically distinct, primarily because the white stripes are Sn-rich, and so the Sn content is much higher in these bead fragments than in the other red beads. A secondary effect is that the blue glass can be seen by the higher Co and As contents.

Among the red beads, the uncored beads (cu, ru, tu) have higher, but not distinct, average Sn contents than the cored or flashed red beads (tcb, tf, bac, cc, rc). This trend is not seen in the Cu contents of the beads, since only the uncored tubular beads tend to be high in Cu. Among the red beads, the uncored tubular beads appear to be chemically distinct. Relative to the other red beads, the uncored tubular beads (tu) are lower in Na, Cl, Ca, and Mn, and are higher in Cu and Sn, as previously observed.

**Black Glasses**

Representative data for the fifty-two, black or grey, round or circular beads are presented in Table VI. The black colouring comes from high (3 % to 7 %) levels of Mn. Tin contents of about 2 % contribute to the greying of some of these black beads. The sole exception to this finding comes from bead 217, which has lower than normal Mn (1.16 %) and is greyed with 2.2 % As. The actual amount of As added as an opacifier is in doubt, since there is 0.26 % of Co in this bead, and the Co may well be associated with 0.5-0.8 % As (see royal blue beads' discussion - above). The remaining fifty-one beads may be sorted by their K contents, to form a low-K group (25 samples- >3% K). Within these two groups, the beads are arranged in Table VI according to their increasing Mn contents.

**Table IV. Representative Red Glass Bead Data Sorted by Bead Shape**

| # | Al % | Ca % | Cl % | Co ppm | Cu % | Mn ppm | K % | Na % | Sn ppm | As ppm | Sb ppm |
|---|---|---|---|---|---|---|---|---|---|---|---|
| 34cu | 1.01 | 5.4 | 0.78 | <73 | 0.78 | 2820 | 3.5 | 9.6 | 23800 | 140 | <48 |
| 26ru | 1.13 | 5.0 | 0.54 | <60 | 1.12 | 4100 | 3.3 | 7.6 | 33400 | 320 | 250 |
| 36ru | 0.99 | 6.5 | 0.71 | <38 | 0.98 | 5940 | 3.3 | 9.4 | 14300 | <45 | <41 |
| | | | | | | | | | | | |
| 142cc | 0.94 | 5.4 | 0.74 | <20 | 0.64 | 1590 | 4.7 | 8.9 | 22700 | 270 | <78 |
| 182rc | 1.01 | 5.3 | 0.82 | <65 | 0.92 | 2990 | 3.9 | 9.5 | 11300 | <53 | <59 |
| 158cc | 0.79 | 5.7 | 0.83 | <68 | 1.03 | 4240 | 3.4 | 9.8 | 3800 | <64 | <69 |
| 141bac | 1.10 | 5.1 | 0.58 | <66 | 1.04 | 4340 | 3.6 | 8.3 | 34400 | 330 | 160 |
| 163cbc | 0.86 | 6.7 | 0.90 | <81 | 0.70 | 4360 | 3.7 | 10.0 | 3300 | <57 | <58 |
| 171rbc | 0.69 | 7.5 | 0.86 | <30 | 0.52 | 4390 | 3.4 | 10.7 | 10200 | <51 | <53 |
| 148bac | 0.82 | 7.1 | 0.85 | <38 | 0.89 | 5600 | 3.8 | 9.5 | 494 | 891 | <43 |
| 164rbc | 0.74 | 6.5 | 0.99 | <34 | 0.54 | 6020 | 2.8 | 10.1 | 5900 | <52 | <56 |
| 169rc | 0.92 | 7.5 | 0.85 | <36 | 0.71 | 7540 | 3.9 | 10.6 | 18000 | <58 | <60 |
| | | | | | | | | | | | |
| 98tu | 0.82 | 4.5 | 0.60 | <21 | 1.09 | 2020 | 2.4 | 7.1 | 35900 | 160 | <37 |
| 101tu | 0.80 | 6.0 | 0.63 | 190 | 0.65 | 3540 | 3.3 | 7.5 | 16700 | 1300 | <44 |
| 97tu | 0.88 | 6.6 | 0.62 | 170 | 1.37 | 6320 | 1.8 | 8.5 | 26000 | 580 | <49 |
| | | | | | | | | | | | |
| 110tbc | 1.09 | 5.2 | 0.73 | <20 | 1.03 | 2430 | 2.9 | 9.1 | 15100 | <64 | <61 |
| 109tbc | 1.02 | 5.0 | 0.60 | <91 | 1.16 | 3210 | 3.3 | 7.9 | 31000 | 340 | <88 |
| 104tbc | 0.68 | 5.2 | 0.62 | <21 | 0.63 | 4700 | 2.9 | 7.9 | 5300 | <58 | <47 |
| 106tbc | 0.66 | 5.4 | 0.89 | <29 | 0.62 | 5810 | 2.6 | 9.6 | 8300 | <60 | <46 |
| | | | | | | | | | | | |
| 117tf | 0.95 | 4.9 | 0.62 | 160 | 1.66 | 2170 | 2.8 | 8.0 | 26900 | 860 | <47 |
| 119tf | 0.61 | 5.8 | 0.64 | <22 | 0.87 | 4220 | 2.5 | 8.2 | 7700 | <38 | <46 |
| 118tf | 0.96 | 5.2 | 0.68 | <74 | 0.96 | 5560 | 2.8 | 9.2 | 14700 | <59 | <47 |
| 127tf | 1.14 | 6.0 | 0.69 | <100 | 1.20 | 6690 | 3.8 | 10.0 | 18400 | <51 | <57 |
| **Bead fragments – wide blue and white stripes on red body.** | | | | | | | | | | | |
| 238 | 0.99 | 5.2 | 0.58 | 130 | 0.98 | 2970 | 3.2 | 7.8 | 57400 | 430 | <110 |
| 241 | 1.03 | 4.5 | 0.55 | 210 | 0.95 | 2880 | 3.6 | 7.7 | 48100 | 510 | <130 |
| 244 | 1.50 | 5.4 | 0.59 | 320 | 1.34 | 3460 | 3.1 | 8.1 | 59900 | 780 | 250 |

| | | | |
|---|---|---|---|
| bac | barrel cored | rc | round cored |
| cbc | circular black cored | rbc | round black cored |
| cu | circular uncored | tf | tubular flashed |
| ru | round uncored | tu | tubular uncored |
| cc | circular cored | tbc | tubular black cored |

**Table V. Means and Standard Deviations for the Differently Shaped Red Beads**

| | cu/ru (12) | bac/cc/ rc/rbc (72) | tu (17) | tbc (13) | tf (13) | b/w/r (8) |
|---|---|---|---|---|---|---|
| Al % | 1.01±0.12 | 0.87±0.14 | 0.86±0.09 | 0.80±0.18 | 0.77±0.21 | 1.17±0.25 |
| Ca % | 6.1±0.6 | 6.3±0.9 | 5.1±0.6 | 5.7±0.5 | 5.6±0.6 | 5.1±0.4 |
| Cl % | 0.71±0.08 | 0.79±0.09 | 0.63±0.07 | 0.71±0.13 | 0.70±0.08 | 0.60±0.04 |
| Co ppm | <70±30 | <60±30 | <50±50 | <40±20 | <70±40 | 270±100 |
| Cu % | 1.00±0.25 | 0.91±0.28 | 1.67±0.59 | 0.96±0.31 | 1.23±0.31 | 1.25±0.21 |
| Mn ppm | 4090±890 | 4840±1080 | 3670±1130 | 4310±810 | 4210±1120 | 3330±300 |
| K % | 3.6±0.5 | 3.5±0.5 | 3.1±0.5 | 3.0±0.7 | 2.9±0.5 | 3.3±0.2 |
| Na % | 9.2±0.8 | 9.6±0.7 | 7.5±0.4 | 8.7±0.7 | 8.7±0.6 | 7.9±0.2 |
| Sn % | 2.09±0.83 | 1.14±0.67 | 2.27±0.92 | 1.50±0.86 | 1.69±0.71 | 5.26±0.63 |
| As ppm | <160±110 | <90±60 | <190±300 | <100±100 | <160±20 | 670±170 |
| Sb ppm | <80±70 | <60±20 | <49±10 | <60±30 | <60±30 | <160±80 |

| | |
|---|---|
| bac | barrel cored |
| b/w/r | wide blue and whites stripes on red body |
| cu | circular uncored |
| ru | round uncored |
| cc | circular cored |
| rc | round cored |
| tf | tubular flashed |
| tu | tubular uncored |
| tbc | tubular black cored |
| rbc | round black cored |

Because of the relatively large amounts of $^{56}$Mn radioactivity generated in these black beads, the analytical sensitivities for elements such as Ca, Cl, Co, Co, and Sn are radically degraded, relative to those found for other colours of glass, so that precisions of measurement are much worse and detection limits are raised significantly. Summary data for the black beads are presented in Table VII. These show that, relative to the high-K group, beads in the low-K group tend to be higher in Ca and Mn, and lower in K and Sn.

The low-K group includes 23 round beads and 2 circular beads. The high-K group contains 10 round beads and 23 circular beads. The existence of two chemical groups with the ten years of operation of the Amsterdam glassmaking house presumably indicates differences in recipe/materials, rather than just corrosion effects.

**Table VI. Representative Round (r) and Circular (c) Black/Grey Glass Beads**

| # | Al % | Ca % | Cl % | Co ppm | Cu % | Mn % | K % | Na % | Sn ppm | As ppm | Sb ppm |
|---|---|---|---|---|---|---|---|---|---|---|---|
| **Low K beads** | | | | | | | | | | | |
| 217c | 1.28 | 7.6 | 0.36 | 2630 | <0.06 | 1.16 | 1.9 | 7.8 | <2300 | 22300 | 620 |
| 45r | 1.38 | 7.1 | 0.49 | <220 | <0.15 | 3.36 | 1.8 | 8.4 | <5900 | 3090 | <47 |
| 209r | 1.03 | 8.3 | 0.68 | <120 | <0.14 | 4.07 | 2.6 | 9.1 | <6000 | 940 | <53 |
| 233c | 1.36 | <1.1 | 0.99 | <370 | <0.14 | 4.38 | 2.3 | 10.8 | <7700 | 500 | <47 |
| 40r | 1.51 | 7.6 | <0.26 | <350 | <0.12 | 5.92 | 1.7 | 7.9 | <6600 | 530 | <79 |
| 232c | 1.41 | 8.9 | <0.34 | <190 | <0.19 | 6.63 | 1.6 | 7.2 | <8700 | 570 | <40 |
| **High K beads** | | | | | | | | | | | |
| 226c | 1.08 | 5.4 | 0.62 | <270 | 0.23 | 2.55 | 4.1 | 8.1 | 13200 | 1430 | <39 |
| 204r | 1.06 | 6.6 | 0.59 | <100 | <0.12 | 3.26 | 3.7 | 7.8 | 21800 | 810 | <48 |
| 200r | 1.03 | 6.1 | 0.53 | <100 | <0.11 | 3.84 | 3.7 | 7.5 | 20700 | 730 | 120 |
| 234c | 1.18 | 6.2 | <0.29 | <160 | <0.16 | 4.90 | 3.9 | 8.0 | 18600 | 590 | <41 |
| 219c | 0.87 | 5.1 | 0.74 | <170 | <0.18 | 6.63 | 3.4 | 9.2 | <8100 | <68 | <56 |

**Table VII. Summary Data for Black/Grey Glass Beads**

| | low K 25 samples | high K 26 samples | bead 217 |
|---|---|---|---|
| Al % | 1.31±0.19 | 1.06±0.09 | 1.28 |
| Ca % | 8.0±1.9 | 5.9±0.9 | 7.6 |
| Cl % | 0.50±0.21 | 0.59±0.16 | 0.36 |
| Co ppm | <250±220 | <210±210 | 2630 |
| Cu % | <0.15±0.03 | <0.19±0.12 | <0.06 |
| Mn % | 4.79±0.93 | 3.95±0.92 | 1.16 |
| K % | <1.6±0.4 | 3.9±0.4 | 1.9 |
| Na % | 8.0±0.9 | 7.9±0.6 | 7.8 |
| Sn % | <0.69±0.16 | <1.68±0.59 | <0.23 |
| As ppm | <730±530 | <770±440 | 22300 |
| Sb ppm | <47±16 | <73±50 | 620 |

"<" implies that detection limit data were folded into the calculation of the mean.

## Colourless Glasses

The twenty-one colourless glass samples come from fragments of glass wasters and flat fragments (see Table VIII). The chemistries of both glass forms are relatively similar, with high Ca (3.8 % to 8.6 %), Na (8.4 % to 11.0 %), and K (2.2 % to 6.0 %), with Mn contents varying over a factor of four (1400 ppm to 6200 ppm).

These are not the major-element glass chemistries of the turquoise blue beads, royal blue beads, or the low-K black beads, but may reflect source glasses for white, red, and high-K black beads.

**Table VIII. Representative Colourless Glasses**

| # | Al % | Ca % | Cl % | Co ppm | Cu % | Mn ppm | K % | Na % | Sn ppm | As ppm | Sb ppm |
|---|---|---|---|---|---|---|---|---|---|---|---|
| **Colourless wasters** | | | | | | | | | | | |
| 279 | 1.16 | 7.1 | 0.69 | <65 | <0.04 | 1440 | 4.1 | 9.8 | <1000 | <45 | <51 |
| 271 | 0.86 | 5.7 | 0.86 | <23 | <0.06 | 2580 | 4.6 | 10.1 | <1000 | <54 | <55 |
| 131 | 1.24 | 4.8 | 0.88 | <24 | <0.05 | 2780 | 3.5 | 11.0 | <1000 | <62 | <49 |
| 273 | 0.60 | 8.3 | 0.97 | <52 | <0.04 | 2850 | 3.0 | 10.0 | <700 | <42 | <47 |
| 287 | 1.33 | 5.8 | 0.84 | <70 | <0.04 | 3660 | 4.1 | 11.1 | <1300 | <62 | <68 |
| 274 | 1.25 | 6.3 | 0.84 | <25 | <0.05 | 4160 | 4.4 | 9.1 | <1100 | <49 | <57 |
| **Cullet scraps - flat sheets** | | | | | | | | | | | |
| 259 | 1.03 | 5.4 | 0.64 | <69 | <0.04 | 1440 | 5.1 | 9.8 | <1200 | <46 | <53 |
| 266 | 0.99 | 6.4 | 0.80 | 58 | 0.05 | 3140 | 5.2 | 9.5 | 1100 | 46 | 51 |
| 267 | 0.88 | 8.6 | 0.96 | <26 | <0.05 | 4740 | 2.2 | 10.5 | 2400 | 140 | <62 |
| 265 | 1.30 | 6.6 | 0.70 | <90 | <0.06 | 5590 | 3.2 | 8.8 | <1900 | <66 | <64 |
| 264 | 0.79 | 5.8 | 0.65 | <35 | <0.05 | 6230 | 5.3 | 8.4 | <1500 | 140 | <38 |

## Gold Glasses

The gold tubular glass samples tended to be covered with flaky black, presumably corrosion-related material that was mostly removable during sonication. Data for the eleven cleaned gold glass samples are presented in Table IX. The primary features of these data are the very low levels of K and Na, and the low levels of Ca and Cl. These features imply that the gold beads are lead-silica glasses.

**Table IX. Data for Gold (Lead) Glasses**

| # | Al % | Ca % | Cl % | Co ppm | Cu % | Mn ppm | K % | Na % | Sn ppm | As ppm | Sb ppm |
|---|---|---|---|---|---|---|---|---|---|---|---|
| 38 | 0.47 | 1.7 | 0.19 | <10 | <0.01 | 601 | <0.07 | 1.8 | 16000 | <8 | <10 |
| 66 | 0.53 | 2.1 | 0.13 | 150 | 0.22 | 3810 | 1.1 | 2.7 | 14100 | 10530 | 700 |
| 70 | 0.43 | 1.4 | 0.17 | 93 | <0.01 | 500 | 0.84 | 1.4 | 15300 | 932 | 7700 |
| 63 | 0.07 | <0.02 | 0.03 | <4 | 0.01 | 59 | 0.015 | 0.017 | 18000 | 119 | 890 |
| 64 | 0.08 | 0.02 | 0.02 | <1 | 0.01 | 18 | 0.019 | 0.014 | 15900 | 63 | 1000 |
| 65 | 0.07 | 0.02 | 0.02 | <3 | 0.01 | 17 | 0.015 | 0.014 | 14600 | 289 | 40 |
| 67 | 0.06 | <0.02 | 0.03 | <4 | 0.01 | 31 | 0.016 | 0.013 | 16700 | 126 | 970 |
| 68 | 0.06 | 0.02 | 0.02 | <3 | 0.01 | 15 | 0.016 | 0.014 | 12900 | 319 | 50 |
| 69 | 0.19 | 0.09 | 0.04 | <5 | 0.11 | 38 | 0.085 | 0.087 | 21200 | 205 | 80 |
| 71 | 0.06 | <0.01 | 0.03 | <3 | 0.01 | 31 | 0.025 | 0.017 | 16100 | 135 | 1160 |
| 72 | 0.06 | 0.03 | 0.03 | <4 | 0.01 | 40 | 0.020 | 0.018 | 15300 | 137 | 1060 |

The eleven samples split into three different coarse chemical groups. The first has only one member, sample 38, with measurable Al, Ca, Cl, Mn and Sn, but no measurable Sb. The second group has two members (samples 66 and 70). This group is similar to sample 38 except that the beads contain about 3 % Sb. The third group has eight members. It is characterised by very low concentrations of all elements except Sn, As, and Sb.

**White Glass Bead Chemical Matches**

Six sets, including four pairs, of white glass samples have similar chemistries (Table X), based on major and trace element concentrations. These sets link the circular beads (c) with white bead fragments (f) and, somewhat surprisingly, with the b/w/r (red) bead fragments. In the case of the b/w/r fragments, they are grouped by their major and minor element contents and are easily discernable from the other glasses by their measurable As (from the Co blue glass) contents. The sets are as follows:

| Set | Samples | Bead form |
|---|---|---|
| 1. | 14, 21, 23, 24, 193, 289 | c - f |
| 2. | 16, 17, 19, 276, 286, 290 | c - f |
| 3. | 250, 196 | b/w/r - f |
| 4. | 246, 281 | b/w/r - f |
| 5. | 13, 275 | c - f |
| 6. | 15, 18. 20, 22, 245, 249, 256 | c - b/w/r; heterogeneous |

**Table X. White Glass Bead Glass Chemistry Groupings**

| # | Al % | Ca % | Cl % | Co ppm | Cu ppm | Mn ppm | K % | Na % | Sn ppm | As ppm | Sb ppm |
|---|---|---|---|---|---|---|---|---|---|---|---|
| 14c | 1.05 | 6.2 | 0.81 | <20 | <0.04 | 2220 | 3.5 | 8.9 | 40700 | <32 | <31 |
| 21c | 0.95 | 5.5 | 0.89 | <55 | <0.04 | 2470 | 3.4 | 8.5 | 47500 | <53 | <46 |
| 23c | 0.99 | 5.9 | 0.92 | <24 | <0.05 | 2270 | 3.6 | 8.9 | 44300 | <42 | <38 |
| 24c | 1.05 | 5.6 | 0.99 | <22 | <0.05 | 2370 | 3.6 | 9.1 | 36900 | <52 | <46 |
| 193f | 0.96 | 5.8 | 0.88 | <21 | <0.04 | 2580 | 4.7 | 9.0 | 51800 | <62 | <58 |
| 289f | 0.94 | 5.7 | 0.95 | <75 | <0.04 | 2450 | 3.9 | 9.0 | 42900 | 110 | <45 |
| | | | | | | | | | | | |
| 16c | 1.02 | 6.0 | 0.92 | <77 | <0.04 | 2900 | 3.7 | 9.4 | 49800 | <69 | <56 |
| 17c | 0.98 | 5.3 | 0.90 | <69 | <0.05 | 2770 | 3.7 | 9.1 | 53800 | <38 | <36 |
| 19c | 0.97 | 6.3 | 0.91 | <25 | <0.04 | 3220 | 3.2 | 8.7 | 46400 | <53 | <53 |
| 276f | 0.99 | 6.4 | 0.91 | <25 | <0.05 | 2780 | 3.9 | 8.7 | 48300 | <39 | <44 |
| 286f | 1.00 | 6.1 | 0.93 | <28 | <0.05 | 2880 | 4.4 | 9.0 | 46200 | 91 | <49 |
| 290f | 1.02 | 6.3 | 0.96 | <33 | <0.05 | 2800 | 3.5 | 9.2 | 51200 | <54 | <62 |
| | | | | | | | | | | | |
| 250bf | 0.89 | 7.1 | 0.75 | 68 | <0.05 | 3650 | 4.5 | 8.8 | 19200 | <65 | <59 |
| 196f | 0.81 | 5.2 | 0.88 | 240 | <0.06 | 3640 | 4.9 | 9.3 | 19900 | 280 | <52 |
| | | | | | | | | | | | |
| 246bf | 0.96 | 5.6 | 0.83 | <78 | <0.06 | 3590 | 4.4 | 9.0 | 28100 | 160 | <59 |
| 281f | 0.91 | 5.8 | 0.67 | 123 | 0.29 | 3590 | 4.2 | 8.9 | 32200 | <76 | <73 |
| | | | | | | | | | | | |
| 13c | 0.99 | 5.3 | 0.75 | <66 | <0.04 | 4340 | 3.9 | 8.6 | 50700 | <52 | <45 |
| 275f | 0.95 | 5.6 | 0.93 | <28 | <0.04 | 4210 | 3.6 | 8.7 | 48800 | <38 | <44 |
| | | | | | | | | | | | |
| 15c | 1.04 | 6.2 | 0.94 | <30 | <0.06 | 5100 | 3.5 | 9.0 | 51000 | <45 | <43 |
| 18c | 0.98 | 6.0 | 1.02 | <32 | <0.06 | 4930 | 4.1 | 9.3 | 14200 | <39 | <37 |
| 20c | 1.00 | 6.6 | 0.89 | <100 | <0.07 | 5480 | 4.8 | 9.3 | 23200 | <48 | <42 |
| 22c | 0.97 | 6.7 | 0.97 | <31 | <0.05 | 5430 | 3.8 | 9.4 | 15300 | <78 | <63 |
| 245bf | 0.83 | 5.6 | 0.94 | 93 | 0.15 | 5210 | 3.8 | 9.2 | 49800 | 200 | <67 |
| 249bf | 0.90 | 5.5 | 0.86 | 110 | 0.15 | 5380 | 3.5 | 9.4 | 39800 | 210 | <59 |
| 256bf | 0.83 | 5.5 | 0.85 | <81 | <0.06 | 5120 | 3.1 | 9.1 | 65900 | 190 | <62 |

c circular white bead
f white bead fragment
b b/w/r fragments

## Chemical Connections with Beads Found in North America

Some example chemistries that illustrate the possibility of tracking glass beads from Amsterdam to North America are presented in Table XI and Table XII. From the white glass beads (Table XI), there appear to be five chemical sets that link the glass beadmaking house in Amsterdam with Seneca sites in New York State. Within the red glass beads (Table XII), the glass beadmaking house is linked by three glass chemistries to Seneca sites; by one chemistry to a Petun site in southern Ontario; and by two chemistries to both Seneca and Petun sites. All of the connected sites are archaeologically dated somewhere within the first two thirds of the 17th century.

**Table XI. Possible Chemical Connections Between White Glass Beads From the Glass Beadmaking House and Seneca Sites in New York State**

| # | Al % | Ca % | Cl % | Cu % | Mn ppm | K % | Na % | Sn ppm |
|---|---|---|---|---|---|---|---|---|
| 25 | 0.56 | 7.0 | 0.61 | <0.02 | 1850 | 3.6 | 7.4 | 68700 |
| ca50 | 0.53 | 5.8 | 0.74 | <0.03 | 1800 | 3.8 | 7.5 | 75500 |
| | | | | | | | | |
| 253 | 0.62 | 7.4 | 0.94 | <0.06 | 2990 | 3.1 | 10.4 | 24000 |
| dh12 | 0.68 | 6.4 | 0.90 | <0.09 | 2990 | 3.2 | 10.0 | 22400 |
| | | | | | | | | |
| 13 | 0.99 | 5.3 | 0.75 | <0.04 | 4340 | 3.9 | 8.6 | 50700 |
| 275 | 0.95 | 5.6 | 0.93 | <0.04 | 4210 | 3.6 | 8.7 | 48800 |
| co37 | 1.02 | 6.2 | 0.86 | <0.04 | 4400 | 4.1 | 9.3 | 54500 |
| | | | | | | | | |
| 20 | 1.00 | 6.6 | 0.89 | <0.07 | 5480 | 4.8 | 9.3 | 23200 |
| 18 | 0.98 | 6.0 | 1.02 | <0.06 | 4930 | 4.1 | 9.3 | 14200 |
| 22 | 0.97 | 6.7 | 0.97 | <0.05 | 5430 | 3.8 | 9.4 | 15300 |
| dh11 | 0.71 | 6.7 | 0.77 | <0.03 | 4940 | 3.5 | 8.9 | 18300 |
| | | | | | | | | |
| 248 | 0.86 | 5.7 | 0.73 | <0.06 | 4560 | 4.1 | 8.2 | 37000 |
| st66 | 0.76 | 6.5 | 0.70 | <0.04 | 4870 | 5.1 | 7.9 | 36100 |

| Code | Site | Period | Region |
|---|---|---|---|
| ca | Cameron | ≈1600-1625 | Seneca |
| dh | Dutch Hollow | ≈1610-1625 | Seneca |
| co | Cornish | ≈1625-1640 | Seneca |
| st | Steeles | ≈1640-1655 | Seneca |

**Table XII. Possible Chemical Connections Between Red Glass Beads From the Glass Beadmaking House and Seneca Sites in New York State, and Petun Sites in southern Ontario**

| # | Al | Ca | Cl | Cu | Mn | K | Na | Sn |
|---|---|---|---|---|---|---|---|---|
| | % | % | % | % | ppm | % | % | ppm |
| **Amsterdam - Seneca connections** | | | | | | | | |
| 30 | 1.07 | 5.0 | 0.65 | 1.03 | 4270 | 4.3 | 8.3 | 27300 |
| 114 | 1.07 | 5.3 | 0.54 | 1.25 | 3930 | 3.5 | 7.5 | 27000 |
| Dh21 | 1.21 | 6.1 | 0.64 | 1.10 | 4290 | 3.3 | 7.9 | 26700 |
| Ph51 | 0.85 | 5.4 | 0.53 | 0.84 | 4240 | 4.4 | 8.0 | 26800 |
| | | | | | | | | |
| 117 | 0.95 | 4.9 | 0.62 | 1.66 | 2170 | 2.8 | 8.0 | 26900 |
| Me46 | 0.97 | 4.9 | 0.67 | 1.98 | 2350 | 2.9 | 8.3 | 25000 |
| | | | | | | | | |
| 28 | 1.10 | 5.7 | 0.75 | 0.85 | 3220 | 3.5 | 9.7 | 23500 |
| Dh01 | 1.02 | 6.6 | 0.72 | 1.17 | 3610 | 3.6 | 9.6 | 20200 |
| Dh16 | 0.98 | 6.1 | 0.71 | 0.83 | 3520 | 2.7 | 9.9 | 24800 |
| **Amsterdam - Petun connections** | | | | | | | | |
| 91 | 0.81 | 4.5 | 0.53 | 2.18 | 4590 | 2.8 | 7.2 | 27500 |
| Pm17 | 0.94 | 4.7 | 0.58 | 1.65 | 4600 | 2.5 | 8.2 | 31800 |
| | | | | | | | | |
| 127 | 1.14 | 6.0 | 0.69 | 1.20 | 6690 | 3.8 | 10.0 | 18400 |
| kc09 | 1.04 | 5.2 | 0.62 | 1.03 | 6950 | 3.2 | 9.9 | 22300 |
| **Amsterdam - Seneca - Petun connection** | | | | | | | | |
| 109 | 1.02 | 5.0 | 0.60 | 1.16 | 3210 | 3.3 | 7.9 | 31000 |
| 31 | 1.15 | 5.5 | 0.60 | 1.03 | 3550 | 3.5 | 8.3 | 32900 |
| Me48 | 0.81 | 5.8 | 0.51 | 1.42 | 3310 | 3.9 | 7.6 | 23000 |
| Pm27 | 0.88 | 5.6 | 0.59 | 1.43 | 3400 | 3.8 | 7.2 | 28500 |
| Da76 | 0.86 | 5.6 | 0.57 | 1.22 | 3630 | 3.5 | 7.4 | 34200 |

| Code | Site | Period | Region |
|---|---|---|---|
| da | Dann | ≈1665-1675 | Seneca |
| dh | Dutch Hollow | ≈1610-1625 | Seneca |
| pm | Plater Martin | ≈1637-1650 | Petun |
| me | Menzis | ≈1640-1655 | Seneca |
| kc | Kelly-Campbell | ≈1639-1649 | Petun |
| ph | Power House | ≈1640-1655 | Seneca |

## Conclusions

Of two hundred and ninety glass samples, all but eleven of leaded, gold-coloured glasses and two potash-rich glasses, were composed of soda-lime-silica glasses. The various chemical groupings found in this sample suite may be used as source elemental fingerprints for the analyses of 17th-century glass beads found in northeastern North America, as well as elsewhere in the world.

The turquoise blue beads were coloured with copper, presumably as Cu(II). The gross glass chemistries of these beads are chemically similar to Ontario Bead Period II (?1600-1620) and Bead Period III (?1620-1650) turquoise blue beads. Cobalt was the colorant for the royal blue beads.

The white glass beads were all coloured with Sn. In white beads with narrow red and dark blue stripes, the Cu (red) and Co (blue) are rarely measurable, although the As, which is associated with Co, is often measurable at low levels. This shows that, for these types of bead, bulk elemental analysis must be used in conjunction with visual inspection of beads and cleaned bead fragments. Chemistries of white glasses spread between bead shapes and glass forms.

The red glass beads were all coloured with Cu, presumably as Cu(I) and/or Cu(0) (e.g. *15*), and opacified with Sn. Cored and uncored red beads appear to be separable based on their Cu and Sn contents, along with differences in the concentrations of other elements.

The broad-blue-on-white-striped, red-bodied beads (b/w/r) are chemically distinct from regular red beads, primarily because the white stripes are Sn-rich, and so the Sn content is much higher in these bead fragments than in the other red beads. A secondary effect is that the blue glass can be seen by the higher Co and As contents.

Black beads are coloured with Mn and sometimes opacified with Sn, forming either black or grey beads. The colourless wasters may reflect source glasses for white, red, and high-K black beads, but are certainly not made of glass chemistries similar to the turquoise blue beads, royal blue beads, or the low-K black beads.

The eleven gold coloured glasses are alkali and alkali-earth poor and therefore must be made of leaded glasses.

## Acknowledgements

This paper is dedicated to the memory of Ian Kenyon who died in 1996. Thanks go to F. Neub (Department of Metallurgy and Materials Science, University of Toronto) for use of his departmental ultrasonic cleaner. A special thanks to K. E. Hancock, who saved the tables from ignominy. This research was made possible by a Social Sciences and Humanities Research Council grant to R.G.V.H.

## References


1. Beck, H. C. *Archaeologia* **1928**, *77*, 1-76.
2. Kidd, K. E.; Kidd, M. A. *Canadian Historic Sites: Occasional Papers in Archaeology and History* **1970**, *1*, 45-89.
3. Karklins, K. In *Glass beads. Studies in Archaeology, Architecture and History*, National Historic Parks and Sites Branch, Parks Canada, Environment Canada. **1985a**; pp 85-123
4. Karklins, K. In *Proceedings of the 1982 Glass Bead Conference*; Hayes, C. F. III, Ed.; Rochester Museum & Scince Center, Research Records No. 16, **1983**; pp 111-126.
5. Hancock, R. G. V.; Chafe, A.; Kenyon, I. *Archaeometry* **1994**, *36*, 253-266.
6. 6. Kenyon, I; Hancock, R. G. V.; Aufreiter, S. *Archaeometry* **1995**, *37*, 323-337.
7. Hancock, R. G. V.; Aufreiter, S.; Moreau J. -F.; Kenyon, I. In *Archaeological Chemistry: Organic, Inorganic and Biochemical Analysis*; Orna, M. V., Ed.; American Chemical Society, Washington, DC, **1996**; Symposium Series 625, pp 23-36.
8. Moreau, J. -F.; Hancock, R. G. V.; Aufreiter, S.; Kenyon, I. In *Proceedings of the Seventh Nordic Conference on the Applications of Science in Archaeology*; Junger, H.; Lavento, M., Eds.; **1997**; 173-181.
9. Hancock, R. G. V; McKechnie, J.; Aufreiter, S.; Karklins, K.; Kapches, M.; Sempowski, M.; Moreau, J. -F.; Kenyon, I. *J. Radioanal. and Nucl. Chem.* **2000**, *244*(3), 567-573.
10. Hancock, R. G. V.; Aufreiter, S.; Kenyon, I. In *Materials Issues in Art and Archaeology*, Materials Research Society, Pittsburg, PA, **1997**; Vol. 462, pp 181-191.
11. Hancock, R. G. V.; Aufreiter, S.; Kenyon, I.; Latta, M. *J. Arch. Sci.*, **1999**, *26*, 907-912
12. Sempowski, M.; Nohe, A.W.; Moreau, J. -F.; Kenyon, I.; Karklins, K.; Aufreiter, S.; Hancock, R. G. V. *J. Radioanal. and Nucl. Chem.* **2000**, *244*(3), 559-566.
13. Thwaites, R.G. (Ed.) *The Jesuit Relations and Allied Documents* (73 vols.); The Burrows Bros. Co., Cleveland, OH; 1896-1901.
14. Karklins, K. *Ornament* **1985b**, *9*(2), 36-41.
15. Freestone, I.C. In *Early vitreous materials*; Bimson, M, and Freestone, I. C., Eds.; Brit. Mus. Occas. Pap., London, UK, **1987**; Vol. *56*, pp 151-164.

## Chapter 9

# Morphology and Microstructure of Marine Silk Fibers

**Rekha Srinivasan[1] and Kathryn A. Jakes[2,*]**

**[1]19 Cenotaph Street, Alwarpet, Chennai, India**
**[2]Department of Consumer and Textile Sciences, Ohio State University, 1787 Neil Avenue, Columbus, OH 43210-1295**

Reports of textiles recovered from marine sites are uncommon and research is needed regarding the physico-chemical structure of water-degraded archaeological textiles and how the marine environment has altered them (*1,2*). Recovery of trunks full of clothing from the ocean floor site of the shipwrecked *S.S. Central America* (*3*) provided a unique opportunity for the study of materials exposed to a marine environment. Recognizing that these textiles could be used to understand both their cultural use in the Gold Rush era and the processes of textile degradation in the deep ocean, research has been conducted on the functions and the structures of the textiles and clothing (*4,5*), as well on the physical and chemical characteristics of the fibers from which the textiles were made (*6-12*). Silk textiles appeared to be in fairly good condition apart from some staining and surface deposits. Morphological features revealed through optical and scanning electron microscopy were compared to those of modern silk and silk obtained from historic textiles of comparable age. Elemental composition of the fibers and surface deposits was determined using energy dispersive spectroscopy. Chemical composition and alterations in secondary structure were studied through infrared microspectroscopy. The implications of the stability of silk in this marine site are discussed.

 

## Marine Textile Research

Reports of textiles recovered from marine sites are uncommon. Wool and flax fibers from the *Mary Rose* and the *Wasa* have been identified employing optical and scanning electron microscopy (*13*). The fiber content of the sails of the *Wasa* were identified as "vegetable" but the method for fiber identification was not described (*14*). Morris and Seifert (*15*) describe treatment of textiles from the *Defence* with oxalic acid but do not indicate how they identified the linen, hemp, and silk fibers that they report are present. Conservation treatment of a silk cocade from the *Michault* was described (*16*).

Peacock (*17,18*) found that aerated soil at pH 6.5 caused more degradation of silk than unaerated soil at a lower pH. She found surface pitting and fibrillation of the fibers as evidence of degradation and suggested that the greater availability of bacteria and fungi in aerated soil results in enzymatic hydrolysis of the silk fibroin. Water-degraded archaeological textiles showed fibrillation, erosion, perpendicular splitting, and pitting. Fibrillation in silk fibers has been attributed to the beginning of very fine axial splitting called lousiness attributed to prolonged wetness, acid hydrolysis, or boiling (*17,18*).

In 1987 Florian (*19*) stated that there was no published data on the analysis of silk from a marine site; the only analysis on silk artifacts recovered from wet or frozen sites reported since then has been that by Peacock in 1996 (*17,18*). No study of the alterations in morphology and chemistry of silk due to exposure to the deep ocean environment has been reported in the literature. Recovery of many silk textiles from the site of the shipwreck of the *S.S. Central America*, therefore, provided an opportunity for examination of the chemical and physical characteristics of silk that had been exposed to a marine environment for an extensive period of time. The steamship sank in 1857, and the trunks were recovered in 1990 and 1991 (*3,6*). Conditions at the site are: depth, 2200 m; solar radiation, nil; pressure, 220 atm; current, 10 cm/s (mean); and visibility, 15 to 20 m (*3*). Additional values anticipated for other parameters were reported as: temperature, 3 °C; dissolved oxygen, 5.8-6.4 mL/L; nitrates, 17-23.9 microgram-atoms/L; silicates, 12-25.9 microgram-atoms/L (*3*). Fibers from the textile artifacts contained within the trunks have been studied using multiple analytical techniques thereby documenting morphological characteristics and elemental composition of deposits (*7,10*). Study of flax fibers from the trunks in comparison to modern flax revealed lower overall crystallinity of the marine material, and increased crystallite size in comparison to modern flax fibers. Unit cell dimensions were reported to be unchanged (*9*).

In an attempt to document the effect of the deep ocean environment on textile materials, modern materials were immersed at the site on the ocean floor for subsequent retrieval. While one set of samples was recovered after a three month period, additional samples have remained on the ocean floor since 1991 and are yet to be retrieved. The morphology of cellulosic fibers immersed for a three-month period has been investigated (*8*). The physical and chemical structure of dyed and undyed cotton fibers from the site compared with those of modern cotton and of cotton immersed on the ocean floor for three months were reported (*10,11*). Results of preliminary analyses on silk fibers using differential scanning calorimetry, energy dispersive x-ray spectoscopy, and scanning electron microscopy, have been reported (*12*).

## Silk Fiber Structure

The filaments of fibroin produced by the larvae of the Bombyx mori (*20*) provide the silk used commercially for textiles. While the specific amino acid composition of fibroin depends on its source, it contains a large percentage of the amino acids glycine, alanine, serine, and tyrosine. The predominance of these simple amino acid residues, which can pack into a compact structure, is one feature of the chemical and physical structure of fibroin that influences performance properties of the filaments (*21*). Small amounts of other amino acids including valine, aspartic acid, glutamic acid, threonine, arginine, leucine, isoleucine, histidine, proline, tryptophan, and cystine are present (*21*). The number of acidic groups on the side chains of this protein is about 2-3 times more than the number of basic groups, affecting such properties as dyeability.

The regularly arranged polypeptide backbone is stabilized by hydrogen bonding between the N-H and the C=O of the amide linkage (*22*). The commonly encountered conformations of the polypeptide backbone include the α-helix, parallel and anti-parallel β-pleated sheets, and random coil conformations (*23*). While the α and β forms are found in crystalline regions of proteins and are characterized by regular arrangement of the polypeptide chain, the random coil conformation corresponds to the amorphous regions and is not characterized by the regular packing of the main chain. A transition phase with laterally ordered regions between the crystalline and amorphous phases has been proposed (*24,25*).

# Experimental Methods

## Samples

To avoid the confounding effect of dye, three textiles (identification numbers 29049, 29054, and 33707) were selected from those recovered marine silks that appeared to be undyed. For comparative purposes, three undyed historic silk textiles (identification numbers 88b, 145b, and 145c) from the period 1860-1880 were obtained from the Historic Costume and Textiles Collection at The Ohio State University. Though the production and usage histories of the marine and historic silks are different, it was assumed that one principal difference between the two sets of silk fibers lies in the fact that the marine silk had been exposed to long term storage in a deep ocean environment, while the historic silks had not. The cumulative effects of 130 years in the two very different environments should be an overriding cause of the most significant differences between the marine and historic silks. Small samples of fabric, ranging in size between 70 and 100 $mm^2$ , were taken from inconspicuous places in the artifacts (e.g., unfinished seam allowances), so as not to affect the appearance of the garments in future display. Modern undyed and unfinished silk (Style #601) obtained from Testfabrics Inc. (Middlesex, NJ), was used in this work as a reference material and is referred to in this document as Reference Silk.

## Optical Microscopy

A Zeiss Axioplan Research Microscope was used to examine the fibers, employing nominal magnifications of 400 X and 1000 X. Between 10 and 15 fibers of each of the samples were mounted onto glass slides using Permount© (Fisher Scientific), a histological mounting medium with a refractive index of 1.55, and allowed to dry for 24 hours before viewing under a microscope. Images were captured using an Optimas 5.2 image analysis system. Diameter measurements were done on 10 fibers per sample at 5 points along each fiber length. After confirming data normality for each sample, one way analysis of variance (ANOVA) was used to test the null hypothesis of no difference between the different sample means. Additionally, Tukey's pairwise comparison procedure was performed to determine the differences between different pairs of means. All statistical analyses were performed using Minitab Statistical software, version 11.0.

### Scanning Electron Microscopy (SEM) and Energy Dispersive Analysis of X-rays (EDS)

In order to investigate fiber morphology, a JEOL JSM 820 scanning electron microscope was used with an accelerating voltage of 20 keV. Analysis of the elemental composition of the fibers and surface contaminants was performed using an Oxford Instruments Group Link Analytical eXL energy dispersive spectrometer with a high resolution Pentafet detector and windowless capability. Fiber samples for both SEM and EDS were mounted on carbon planchettes with carbon tape and sputter-coated with carbon using a Denton Vacuum Desk II coater. Between 5 and 10 individual fibers of each sample were mounted for examination.

SEM photographs were taken under a variety of magnifications in order to elucidate various morphological characteristics. When the fibers themselves were being analyzed using EDS, spectra were collected at a magnification of <1000 X so as to obtain the average elemental composition of the entire sampling area. On the other hand, in analyzing surface encrustations, a more focussed spot sampling of the specific characteristic was performed at higher magnifications.

Features of specimen morphology noted included smooth fibers, longitudinal striations, pits on fiber surface, fibrils, surface cracks, deep cracks, mechanical fiber deformation, fiber flattening, and the formation of fibrils. Longitudinal striations were defined as ridgelike surface characteristics in the direction of the fiber axis. The spacing between these striations was qualitatively estimated as being regular or irregular, based on the distance between the ridges. Surface pitting phenomena refer to areas on the fiber that appear to have been eaten away. Cracks were classified as being either surface cracks, i.e. those restricted to the fiber surface, or deep cracks, those that penetrate the fiber bulk. Mechanically deformed fibers appeared to have been twisted and distorted.

In addition, deposits on the surface of the fibers were examined. Based on the computer generated elemental composition data obtained from the EDS analysis, the arbitrary concentration definitions suggested by Goldstein et al. (*26*) were used to classify constituents: major (>10 weight %), minor (1-10 weight %), and trace (<1 weight %).

### Fourier Transform Infrared Microspectroscopy (FTIR)

Chemical composition of the fibers was analyzed using a Bruker Equinox 55 IR spectrophotometer, equipped with a Mercury Cadmium Telluride detector and a nitrogen purged microscope attachment. The instrument operation parameters were as follows: 4.0 $cm^{-1}$ resolution, 520 scans per spectrum, Blackman-Harris apodization function, zero-filling factor of 2, and the gain switch turned on. Data collection was done using a magnification of 15x and an aperture size of 0.3 mm, which resulted in a spot size of about 20 μm. Bruker's OPUS IR spectroscopic software version 2.2 was used for data collection and data analysis was performed using Grams 386 (Galactic Industries) software.

Fifteen to 20 single fibers, each approximately 15mm in length, were mounted to lay across a 4mm by 20mm slit in a heavy-duty cardboard frame. The fibers were held in place with double-sided tape. The portions of the fiber in the slit were gently flattened using a roller. Flattening was required to ensure that the sample covered the aperture and to minimize diffraction effects. This method was chosen over other flattening techniques like using a KBr window or a diamond cell, because it was considered less disruptive to fiber internal structure than those techniques.

For each specimen 50 spectra were collected and analyzed. These spectra were collected on 10 different fibers for each specimen, by sampling 5 spots along each fiber. Care was taken to ensure that spectra were collected in spots that were relatively free of surface encrustations, and that fibers of varying thickness were sampled. As the technique is nondestructive, and the sample size small, five fibers from each specimen type were sampled, and five spectra were collected at different locations along the length of each fiber.

Similarity of spectra collected from a single fiber was visually estimated by superimposing the spectra. In order to estimate the intra-specimen spectral variation, average and standard deviation spectra were computed over the 50 spectra for each specimen. Both standard peak picking and peak picking using the second derivative spectra were employed. The standard peak picking method evaluates the x-values of the interpolated maxima of the absorbance spectra. An absorbance threshold of 2% was used in this technique. Peak picking using the second derivative method evaluates absorbance maxima from peak minima in the second derivative spectra. This is particularly useful in the identification of strongly overlapping bands and weak shoulder bands. A threshold of 3% and a 9 data point smoothing value was used. Peaks were assigned to atomic motions of the main chain bonds, the highly localized side chains of the protein amino acids, and to specific amino acid sequences based on reported literature. The presence or absence of these bands was qualitatively compared across specimens.

The ratio of the absorbance intensities of the infrared bands at 1265 and 1235 $cm^{-1}$ is commonly used in fibroin crystallinity estimations (*27,28,29*). The band at 1235 $cm^{-1}$ is assigned to the Amide III mode of the random coil conformation, and the corresponding band of the β-sheet conformation has been identified at 1265 $cm^{-1}$ (*30*). Since these two peaks were not well resolved in the spectra under consideration, intensity measurement of the two peaks was difficult. In order to alleviate this problem, a Fourier self-deconvolution was performed in the 1800-800 $cm^{-1}$ region of the spectrum, which aided in resolving the two broad peaks into narrower ones. The boundaries of the range to be deconvoluted were chosen so that the intensities at the boundaries were close to zero (*31*). Deconvolution assumes that the measured spectrum is comprised of a series of narrow lines convolved by a common line-broadening factor (LBF) whose effect can be mathematically removed by multiplying the spectrum with a deconvolution function which is the inverse Fourier transform of the LBF.

A Lorentzian LBF was used in the deconvolution. A deconvolution factor of 100 was chosen in order to achieve a net amplification of 3.4. This value was chosen after trials with deconvolution factors of 50, 100 and 1000 in the manner recommended by Kauppinen et al. (*32*). With a factor of 50, under-deconvolution is observed, while a factor of 1000 results in over-deconvolution and negative side lobes. The deconvolution function amplifies the noise as well, and in order to reduce the noise, apodisation with a Blackman function is performed in conjunction with the deconvolution over the fraction of the interferogram specified by a noise reduction factor. A noise reduction factor of 0.5 was chosen. While the intensities of the deconvoluted peaks are higher than those in the original spectra, the relative peak intensities in the deconvoluted spectra remain the same as in the original (*31*).

Relative intensities of the peaks at 1235 and 1265$cm^{-1}$ were estimated by drawing a local baseline and measuring the peak heights on the deconvoluted spectra. The crystallinity ratio $A_{1265cm^{-1}}/A_{1235cm^{-1}}$ was computed for each spectrum, and the average and standard deviation computed for each specimen. After confirming data normality for each sample, one way analysis of variance (ANOVA) was used to test the null hypothesis of no difference between the different sample means. Additionally, Tukey's pairwise comparison procedure was performed to determine the differences between different pairs of means. All statistical analyses were performed using Minitab Statistical software, version 11.0.

## Results and Discussion

### Morphology and Elemental Composition

The average and standard deviation of the fiber diameters in the Reference, Historic, and Marine Silk specimens are summarized in Table I. Analysis of variance of the diameter data indicates that there is a significant difference in fiber diameters among the silk groups (p = 0.00). Based on Tukey's pairwise comparisons, the fiber diameters of the Marine Silk 29049 and 33707 do not differ significantly from each other and they are significantly larger than the Reference silk, Historic Silks 145b and 145c, and Marine Silk 29054.

**Table I. Average and Standard Deviation of Fiber Diameter Measurements for the Reference Marine and Historic Silk Specimens**

| *Specimen ID* | *Average (μm)* | *Standard deviation (μm)* |
|---|---|---|
| Reference | 13.383 | 1.103 |
| Marine 29049 | 14.035 | 1.172 |
| Marine 29054 | 13.386 | 1.177 |
| Marine 33707 | 14.048 | 1.140 |
| Historic 88b | 13.403 | 1.030 |
| Historic 145b | 13.368 | 1.060 |
| Historic 145c | 13.347 | 0.996 |

As observed with SEM, the fiber surfaces of the Reference Silk are smooth and rounded and show none of the signs of degradation or morphological alteration displayed by the Historic and Marine Silk fibers. Fiber surfaces are clean of external deposits or encrustations, and show no longitudinal striations. The surface of the majority of the Historic Silk fibers are smooth and rounded, resembling the Reference silk. Occasionally however, the Historic Silk fibers do reveal some alterations in morphology including surface cracking and surface pitting. A small amount of surface deposit is seen on Historic specimens 88b and 145c. This deposit appears to be selective and is restricted to a few fiber surfaces (Figure 1). These deposits appear flat-faceted and crystalline; their elemental composition is summarized in Table II.

In comparison with the other two sample types described above, the marine fibers reveal the most alteration in fiber morphology. However, the characteristics are not present uniformly across the different Marine samples, or even in all areas of the same sample. While parts of the specimens examined appear very similar to the Reference Silk, some areas reveal considerable morphological alteration and degradation, along with extensive deposits on the fiber surface. A few fibers in all three Marine samples show signs of severe mechanical deformation (Figure 2) while others appear twisted or flattened. Longitudinal striations are observed along the fiber length in all three Marine Silk specimens. These striations sometimes appear to have regular spacing between them (Figure 3), while at other times the striations are irregularly spaced.

*Figure 1. Historic Silk 145c: surface deposits.*

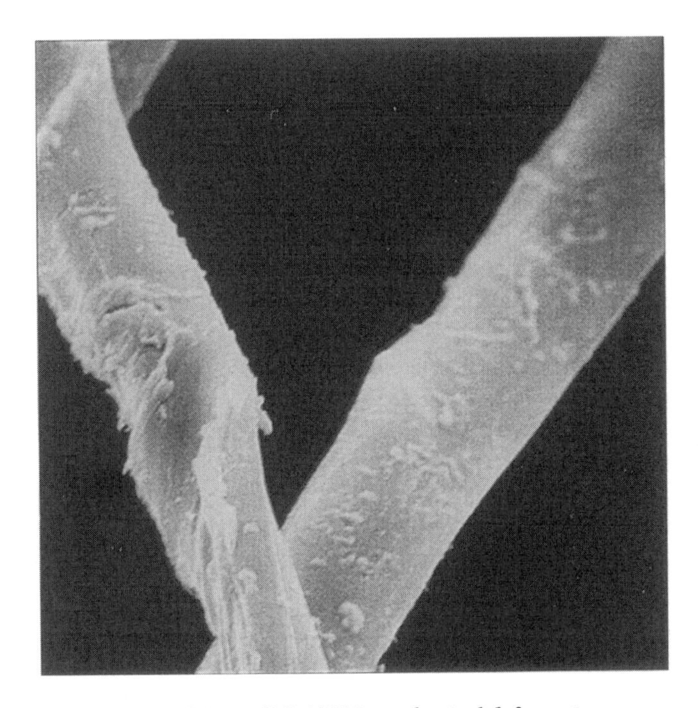

*Figure 2. Marine Silk 29054: mechanical deformation.*

Two different cracking patterns are observed in the Marine specimens. The first type of cracking, which appears restricted to the surface of the fibers, is observed only in Marine Silk 29054. It is characterized by vertical and horizontal cracks with respect to the fiber axis. As this feature was observed only occasionally, no quantitative estimation of the crack spacing was made. The second type of cracking penetrates deeper into the fiber bulk. This type of cracking is observed in two areas of the specimens. First, this type of cracking is seen where fiber twisting occurs and appears to be a consequence of the twisting. This is observed in Marine Silk 29049 and 33707 (e.g. Figure 4). These cracks are approximately parallel to the fiber axis and are of varying lengths between 1-15 μm. The second area where this cracking phenomenon is observed in all three Marine samples is in areas that are characterized by heavy surface deposits. One feature observed in Marine Silk that is not seen in Historic specimens is the occurrence of fibrillation. While moderate amounts of this characteristic are observed in Marine Silk 33707, this feature is seen only occasionally in Marine Silk 29049 and Marine Silk 29054.

The extent of extraneous deposits on the Marine specimens varies widely. Some areas are almost completely free of deposits, others have moderate amounts of deposits, and some areas are so heavily covered that the fiber surface itself is not visible. Of the Marine Silks, 29049 and 33707 have the most deposits, and 29054 has fewer deposits. In comparison with the Historic Silks however, all Marine Silks have more deposits. These deposits appear very different from those observed on the Historic Silks. The first is a continuous pastelike deposit with small beaded structures less than 0.5 μm in diameter on its surface (Figure 5). These deposits are similar to those reported by Chen and Jakes (*10*) and Jakes and Wang (*7*) on cellulosic fibers from the *SS Central America*. Second, discrete cube shaped particles of varying size up to 1.5 μm in diameter that appear crystalline (Figure 6). Similar particles were found on cotton fibers from the same site (*10*). Third, irregular shaped discrete particles approximately 2 to 2.5 μm in diameter were observed occasionally. Table II summarizes the elemental composition of the cube shaped discrete deposits, irregular shaped deposits, and continuous pastelike encrustations observed on the marine fibers.

The continuous pastelike deposit is not the same as the other deposits. It contains no elements larger than sodium, and may be organic. It is likely that these deposits are the result of precipitation of suspended particulate and colloidal matter that forms a sedimentary deposit over the ocean floor (*19*).

The irregular discrete deposits observed on the Marine Silks, composed primarily of Cl, Fe, Na, and S differ from the cubic encrustations in their high

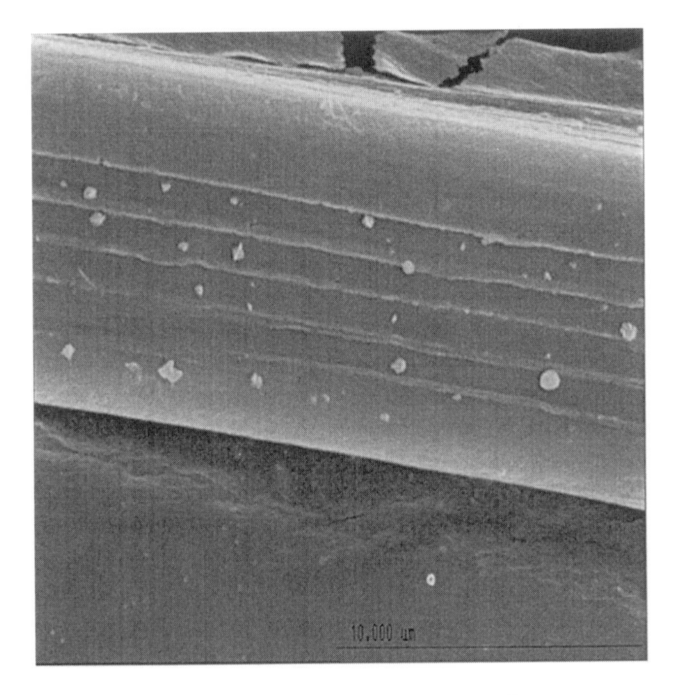

*Figure 3. Marine Silk 33707: regularly spaced striations.*

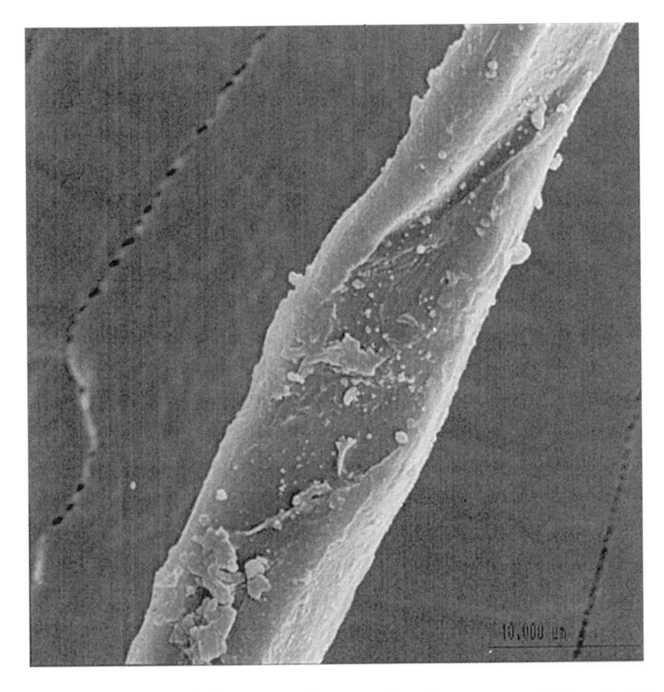

*Figure 4. Marine Silk 33707: deep cracks along areas of fiber twisting.*

**Table II. Elemental Composition of Surface Deposits and Encrustations Observed on Historic and Marine Silk Fibers**

| *Element* | *Specimen/Morphological Characteristics* | | | | | |
|---|---|---|---|---|---|---|
| | Historic Silk 88b:deposit | Historic Silk 145c:deposit | Marine Silk 29049:cubic deposit | Marine Silk 29054: cubic deposit | Marine Silk 33707; irregular deposit | Marine Silk 33707: continuous paste deposit |
| Al | L | M | S | S | S | |
| Ca | M | | | | | |
| Cl | S | S | S | S | L | |
| Fe | M | | M | | L | |
| K | M | | S | S | | |
| Mg | M | M | L | L | M | |
| Na | L | L | L | L | L | |
| P | | S | | | S | |
| S | M | L | L | L | L | |
| Si | L | L | M | S | | |
| Sn | | | | | S | |
| Zn | | | L | L | | |

L indicates a major constituent; M indicates a minor constituent; S indicates a trace constituent

concentration of Fe, and the fact that they do not contain significant amounts of Zn. It is likely that these are the result of the activity of sulfacte reducing bactera active within the trunk. Hydrogen sulfide is produced as a result of microbial activity, which subsequently reacts with dissolved ferric and ferrous species to produce iron sulfides. Iron sulfide residues are found extensively on marine artifacts including pottery and wood (*19*). In contrast to degradation observed in cotton, however, no evidence of biodeterioration of the silk fibers was observed and no microorganisms were observed on the silk fiber surfaces.

The cubic deposits observed in Marine Silk 29049 and 29054 contain Na, Mg, S, and Zn, with Al, Cl, K, and Si in smaller proportions. The difference

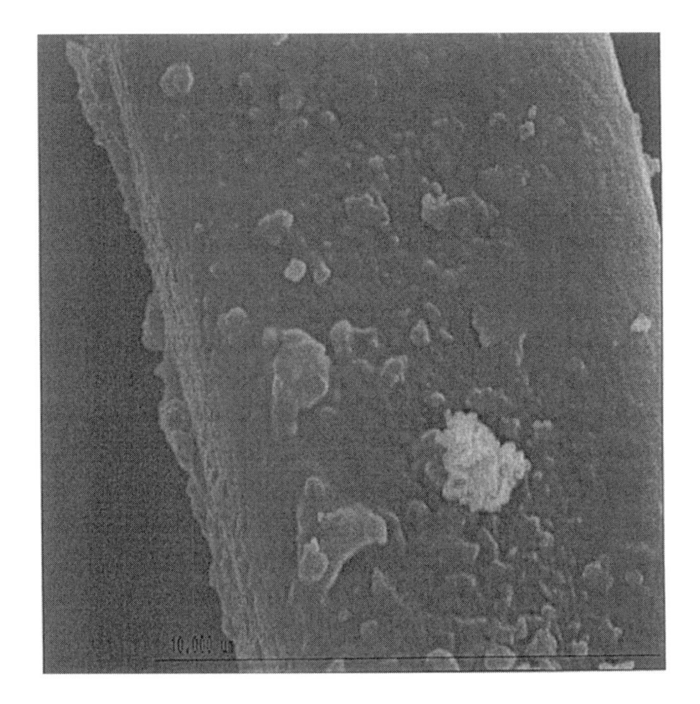

*Figure 5. Marine Silk 33707: continuous paste-like deposit.*

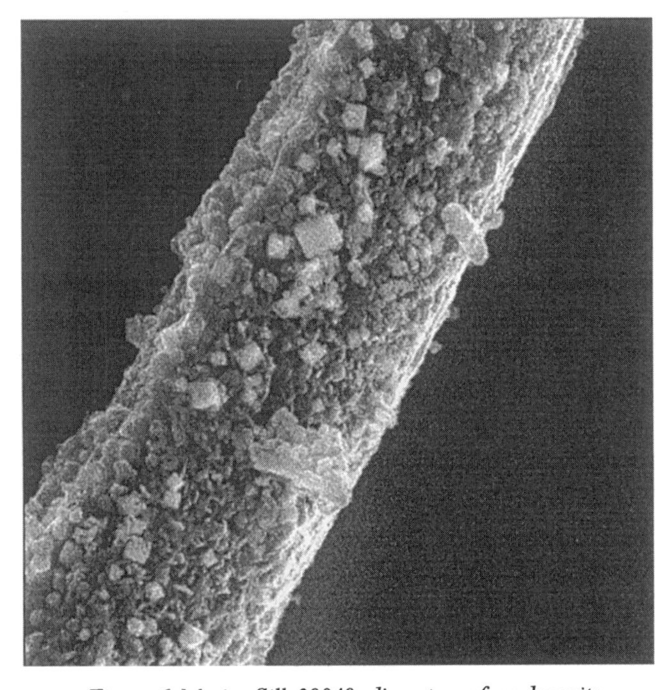

*Figure 6. Marine Silk 29049: discrete surface deposits.*

between these two specimens is that while 29049 contains Fe as a minor constituent, 29054 does not contain this element. These deposits contain elements from the ocean water that have precipitated on the fiber surfaces. The deposition of siliceous material, for example, is possible when silica in solution encounters a surface of pH lower than 8.5 (*19*). The pH of the trunk contents is altered as a consequence of the action of the sulfate reducing bacteria, altering the solubility of some of the dissolved components of the ocean water contained therein.

## Chemical Composition

Infrared spectra of spots collected along the same fiber show little or no variation, both with respect to band positions and intensities. When spectra from different fibers of any given specimen were visually compared, they showed some variation with respect to peak intensities. Peak positions were fairly consistent across spectra for each specimen, though some variation in peak sharpness was noted. The peak sharpness was especially variable in the peak at 3298 $cm^{-1}$. In order to estimate the intra-specimen spectral variation, average and standard deviation spectra were computed over the 50 spectra for each specimen. Figure 7 shows an example of the average and standard deviation spectra obtained for Marine Silk 29049. The standard deviation for all samples is fairly low, indicating that spectral variability within a specimen is small. However, the extent of deviation varies across the spectral range, and is higher in specific areas of the spectra. In particular, the 3400-3200 $cm^{-1}$ and 1700-1500 $cm^{-1}$ wavenumber ranges show maximum variability.

Comparing the standard deviation spectra of the specimens indicates that two of the Marine Silk artifacts, 29054 and 33707, exhibit more variability than the others particularly in the 1500-880 $cm^{-1}$ range. The third marine specimen, 29049, and all three Historic Silks, 145b, 145b, and 88b, exhibit the least variability across fibers. In these samples, the most variability was observed in the 3400-3200 $cm^{-1}$ region, with little variability across the rest of the spectral range.

All peaks identified in the standard peak picking technique were also identified with the second derivative method but the latter identified some additional peaks. Figures 8a, 8b, and 8c show the second derivative spectra of the Marine specimens in comparison to the Reference Silk spectra. Table III summarizes the results of both the standard and the second derivative peak picking methods for the silk fibers. All peaks were consistently present in all three groups of spectra, and peak positions in the spectra of the Historic and Marine fibers varied from the peak positions in the Reference fibers over a ±10 $cm^{-1}$ wavenumber range. The difference in peak parameters between the spectra

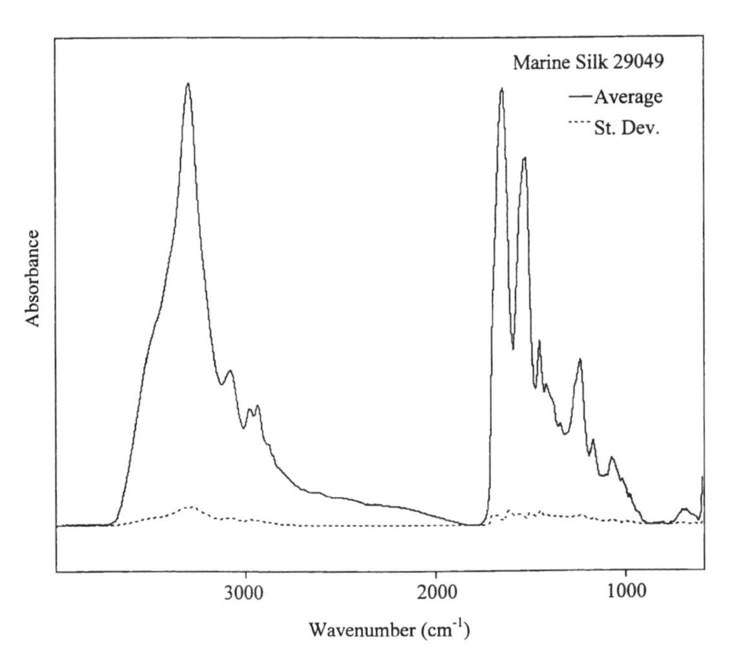


*Figure 7. Marine Silk 29049: average and standard deviation absorbance spectra.*

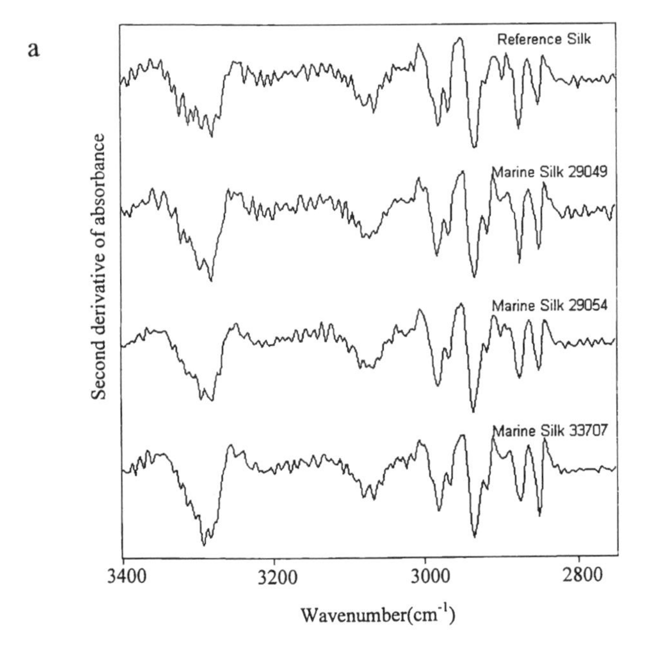


*Figure 8. Reference and Marine silks: Second derivative spectra. a. 3400 $cm^{-1}$ to 2600 $cm^{-1}$. Continued on next page.*

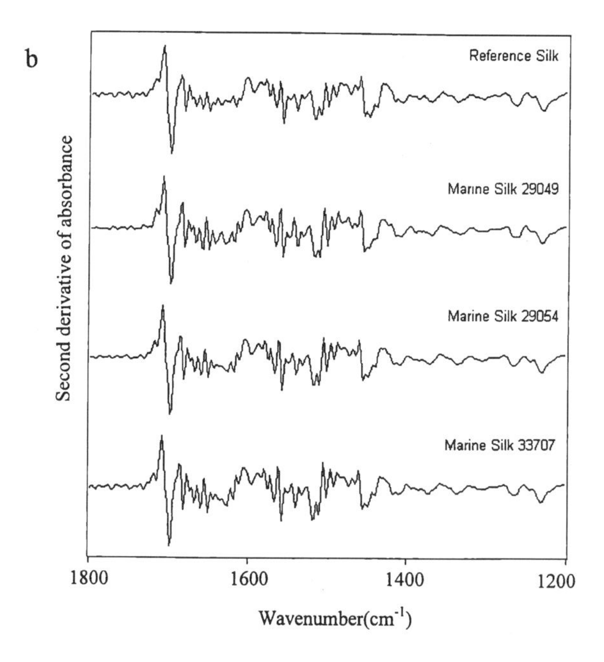


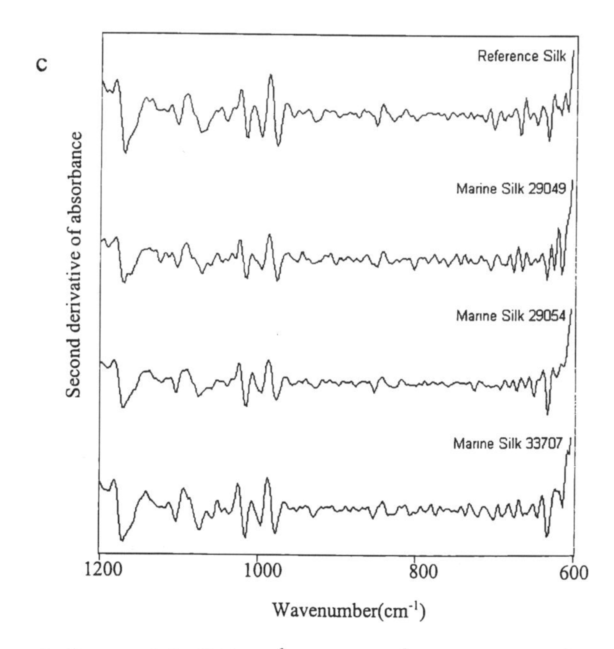


*Figure 8. Continued. b. 1800 $cm^{-1}$ to 1200 $cm^{-1}$ and c. 1200 $cm^{-1}$ to 600 $cm^{-1}$.*

**Table III. Infrared Peak Identification and Assignment Using Standard and Second Derivative Peak Picking Techniques.**

| *Band assignment* | *Reference silk* | | *Marine Silk 29049* | | *Marine silk 29054* | | *Marine Silk 33707* | | *Historic Silk 88b* | | *Historic Silk 145b* | | *Historic Silk 145c* | |
|---|---|---|---|---|---|---|---|---|---|---|---|---|---|---|
| | S | SD | S | SD | S | SD | S | SD | S | SD | S | SD | S | SD |
| NH stretching β sheet | 3296 | 3286 | 3296 | 3284 | 3296 | 3284 | 3295 | 3292 | 3296 | 3286 | 3296 | 3286 | 3296 | 3286 |
| NH stretching β sheet | 3080 | 3079 | 3081 | 3076 | 3083 | 3072 | 3082 | 3080 | 3079 | 3080 | 3080 | 3075 | 3080 | 3077 |
| CH stretching | 2979 | 2981 | 2980 | 2983 | 2980 | 2983 | 2979 | 2982 | 2979 | 2982 | 2980 | 2983 | 2980 | 2983 |
| CH stretching | 2933 | 2934 | 2935 | 2935 | 2935 | 2935 | 2935 | 2935 | 2934 | 2933 | 2936 | 2935 | 2935 | 2935 |
| CH stretching | | 2875 | | 2875 | | 2976 | | 2875 | | 2975 | | 2876 | | 2876 |
| Amide I random coil | | 1700 | | 1699 | | 1699 | | 1699 | | 1700 | | 1699 | | 1699 |
| Amide I β sheet | | 1663 | | 1663 | | 1663 | | 1663 | | 1663 | | 1663 | | 1663 |
| | 1648 | | 1644 | | 1644 | | 1645 | | 1645 | | 1646 | | 1645 | |
| Amide I β sheet | | 1633 | | 1629 | | 1629 | | 1630 | | 1631 | | 1630 | | 1630 |
| Amide II β sheet | | 1555 | | 1553 | | 1554 | | 1554 | | 1555 | | 1554 | | 1554 |
| Amide II random coil | | 1539 | | 1535 | | 1536 | | 1537 | | 1538 | | 1537 | | 1537 |
| | 1519 | | 1519 | | 1520 | | 1520 | | 1519 | | 1520 | | 1520 | |

| Mode of p-disubstituted benzene ring (tyrosine) | | 1514 | | 1514 | | 1514 | | 1514 | | 1514 | | 1514 | | 1514 |
|---|---|---|---|---|---|---|---|---|---|---|---|---|---|---|
| CH3 modes of alanine | 1448 | 1450 | 1449 | 1450 | 1449 | 1451 | 1449 | 1450 | 1448 | 1450 | 1449 | 1451 | 1449 | 1451 |
| CH deformation | 1410 | 1409 | 1411 | 1412 | 1412 | 1412 | 1412 | 1411 | 1410 | 1409 | 1411 | 1413 | 1411 | 1413 |
| Amide III β-sheet | | 1266 | | 1265 | | 1266 | | 1265 | | 1265 | | 1265 | | 1265 |
| Amide III random coil | 1233 | 1232 | 1234 | 1232 | 1234 | 1232 | 1234 | 1232 | 1232 | 1231 | 1234 | 1232 | 1234 | 1232 |
| CH3 modes of alanine | 1168 | 1169 | 1168 | 1169 | 1168 | 1170 | 1168 | 1169 | 1167 | 1169 | 1168 | 1170 | 1168 | 1170 |
| | 1069 | 1071 | 1070 | 1072 | 1070 | 1973 | 1070 | 1073 | | 1072 | 1069 | 1072 | 1068 | 1072 |
| Ala-gly and gly-ala | | 997 | | 997 | | 997 | | 997 | | 997 | | 996 | | 996 |
| Ala-gly and gly-ala | | 975 | | 975 | | 975 | | 976 | 976 | 975 | | 975 | | 975 |
| Amide IV β-sheet | 701 | 701 | 698 | 700 | 697 | 698 | 700 | 702 | 701 | 697 | 700 | 701 | 700 | 701 |
| Amide IV random coil | | 634 | | 635 | | 633 | | 632 | | 634 | | 634 | | 634 |

S = standard spectrum; SD = second derivative spectrum

collected in this work and the peaks identified for silk fibroin in the literature was about ±5 $cm^{-1}$ wavenumbers.

Carbonyl formation due to oxidation results in an infrared band at 1735 $cm^{-1}$ in silk fibroin (*33*) but no evidence of this band was found in the Historic and Marine Silks. Degradation of fibroin both as a result of exposure to light (*34,35,36*) and chemical oxidants (*37,38,39*) has been shown to notably alter the tyrosine content of fibroin. However no evidence of alteration of the infrared band at 1515 $cm^{-1}$ attributed to the tyrosine moiety (*30*) was found in the Historic and Marine Silks in comparison to the Reference Silk.

The average crystallinity ratios calculated from the intensities of the 1265 and 1235 $cm^{-1}$ absorbances are summarized in Table IV. The crystallinity ratio determined for the Reference Silk, 65%, compares closely to the values reported in the literature (*27,28,29*). A one way ANOVA indicates that there are significant differences between the crystallinity ratios of the silk specimens. Based on Tukey's pairwise comparisons, the crystallinity ratios of the Reference Silk and of Historic Silk 88b are significantly larger than the other two Historic Silks and the Marine Silks. The smaller crystallinity index of these latter materials is indicative of a change in short-range order, a disruption of the local environment of the atoms responsible for the infrared peaks. Thus, while the infrared spectra of the silk specimens are comparable and indicative of similarity in their chemical composition, the ratio of peaks associated with random coil and β-sheet conformations has changed, indicative of a change in ordering of the molecules relative to each other.

## Discussion

Increased fiber diameters of the Marine Silks compared to the Reference Silk is indicative of a disruption in molecular association that occurred in the ocean and is maintained upon drying from the waterlogged state. Upon immersion in water, inter- and intra-sheet protein-protein hydrogen bonds are replaced by water-protein bonds in the amorphous regions, leaving weaker van der Waals forces as the dominant noncovalent protein-protein interaction (*40*). An inverse relationship between filament size and molecular orientation has been demonstrated in the literature (*41*); similarly the larger diameter of these silk fibers is indicative of internal structural change.

**Table IV. Average and Standard Deviation of Infrared Crystallinity Ratios**

| *Specimen ID* | *Average* | *Standard Deviation* |
|---|---|---|
| Reference | 0.65215 | 0.07993 |
| Marine 29049 | 0.59240 | 0.07559 |
| Marine 29054 | 0.56033 | 0.06825 |
| Marine 33707 | 0.57307 | 0.07196 |
| Historic 88b | 0.62593 | 0.06862 |
| Historic 145b | 0.57243 | 0.07283 |
| Historic 145c | 0.58227 | 0.07291 |

While significant morphological alteration and increased fiber diameters are noted in the Marine Silks, infrared spectroscopic evidence suggests that neither the age of the Historic Silks nor the combined effect of age and the deep ocean environment appear to have chemically altered the structure of the silk fibroin. The chemical stability of these materials can be attributed to a number of factors. First, all specimens came from relatively unexposed areas of the artifacts, i.e. from seam allowances, and should therefore have been the least degraded of any area of the garment prior to immersion in the deep ocean. Chemical and photooxidation occur maximally at exposed surfaces of materials (*42*), and it quite likely that limited exposure to surrounding chemical agents during the lifetime of the garment protected these areas of the fabric. The absence of weighting agents on the selected silks could also play a part in their chemical stability; weighting agents such as tin, lead, and zinc are known to exacerbate protein degradation.

Upon submersion in the trunk on the ocean floor, the extremely cold temperature and the lack of light within the trunks would inhibit deterioration of the marine silk artifacts. In addition, while the pH of the marine environment at the shipwreck site ranges from 7.99 to 8.14 and the oxygen concentration is 5.85-6.40 mL/L (*3*), the confined conditions within the trunks result in an anoxic environment with a somewhat reduced pH. As oxygen is consumed within the trunk by the action of aerobic microbes decomposing organic material, the redox potential would be lowered (*43*) and silk deterioration would become limited (*44*). As the environment becomes anoxic, sulfate reducing bacteria would thrive, producing sulfides that in turn cause an increase in acidity (*19*). Consequently, in the near vicinity of the silk artifacts the conditions could readily fall within the isoionic range of silk, i.e., in the range between pH 4 to 8. In this region, the degradation of silk due to acidic or alkaline hydrolysis is minimized.

Because silk textiles can persist in a marine environment for a considerable period of time, the marine archaeologist should be aware that it is likely that silk textiles submerged with other shipwrecks has survived as well. Appropriate care upon recovery, particularly maintaining their wet state in darkness and in the same ocean water will protect them until they can be properly assessed in the laboratory.

Beyond the typical actions a conservator may take with respect to inhibition of degradation of material artifacts, such as controlled temperature, humidity, and light, certain characteristics of Marine Silk should be considered in determining conservation treatments. Marine Silks have a more accessible structure as evidenced by their lower crystallinity and larger fiber diameter in comparison with Reference Silk. Additional evidence of alteration to the physical structure of Marine Silk comes from striations, cracking, and flattening phenomena observed in these artifacts. The extensive mineral deposits on the surface of Marine Silks could result in an increased rate of degradation of the textiles upon recovery into a terrestrial environment. While vacuuming with a low powered vacuum unit may remove some of the surface contaminants, cleaning with a chelating agent is likely to be necessary. Wet cleaning of textile artifacts with deionized water alone has been shown to successfully alter the pH of such artifacts (*45*). Wet cleaning is suggested for Marine Silk artifacts to remove the residual salts, remove surface deposits, and lower the pH of the artifacts to bring them into a stable isoionic range. For example, the use of a mercaptoacetic acid bath to remove iron compounds on the surface of silk textiles from the *Machault* has been reported to be successful (*16*). Prior to using such a technique however, systematic testing of the treatment on small samples should be conducted.

## Conclusion

In comparison to Reference Silk and Historic Silk fibers, the Marine Silk fibers show more extensive morphological damage with deep cracks, surface striations, fibrillation, and fiber flattening. These changes in morphology, combined with the increased diameter of two of the Marine Silks, reflect changes in the physical microstructure of the artifacts as a result of the deep ocean environment.

None of the Marine Silks display significant inorganic content, but they are encrusted with both inorganic and organic deposits that reflect the anoxic environment established within the trunk. All of the infrared absorbance peaks characteristic of fibroin were identified in the Reference, Historic, and Marine Silks and no new peaks were identified in either the Historic or the Marine specimens. Thus the silk fibers appear to be unchanged by their long term exposure to the marine environment. Several factors of the environment within the trunks, including reduced circulation, oxygen depletion, low temperature, darkness, and pH in the isoionic range of fibroin act together to promote chemical preservation of the artifacts. However, while the protein structure appears to be chemically unchanged, the infrared crystallinity ratios of the marine silks are indicative of a change in the short range order of the protein molecules. The implications of these findings include the need for marine archaeologists to be aware of the possibility of the discovery of silk at marine sites, and the need for textile conservators to be aware of the potential for marine silks to react differently to treatment than silks from terrestrial sites.

## References


1 Peacock, E.E. In *Biodeterioration and Biodegradation*; Rossmore, H. W., Ed.; Elsevier Applied Science: New York, 1990; pp. 438-439.
2 Pearson, C. In *Conservation of Marine Archaeological Objects*; Pearson, C. Ed.; Butterworths: London, 1987; preface.
3 Herdendorf, C. E.; Thompson, T. G.; Evans, R. D. *Ohio J. Sci.* ***1995**, 95*, 4-224.
4 Crawford, L. C. Ph.D. Thesis, The Ohio State University, Columbus, 1994.
5 Hannel, S. L. Master's thesis. The Ohio State University: Columbus, 1994.
6 Jakes, K. A.; Mitchell, J. C. *J. Am. Inst. Conserv.* **1992**, *31*, 343-353.
7 Jakes, K. A.; Wang, W. *Ars Textrina* **1993**, *19*,161-183.
8 Wang, W.; Jakes, K. A. *Ars Textrina* **1994**, *22*, 33-64.
9 Foreman, D. W.; Jakes, K.A. *Text. Res. J.* **1993**, *63*, 455-464.
10 Chen, R.; Jakes, K. A. *J. Am. Inst. Conserv.* **2001**, *40*, 91-103.
11 Chen, R.; Jakes, K. A. In *Historic Textiles, Paper, and Polymers in Museums*; Cardamone, J. C.Ed.; American Chemical Society: Washington, D.C., 2001; pp. 38-54.
12 Srinivasan, R.; Jakes, K. A. In *Materials Issues in Art and Archaeology V. Mater.Res.Soc.Symp.Proc.* 462; Vandiver, P. B; Druzik, J. R.; Merkel, J. F.; Stewart, J. Eds.; Materials Research Society: Pittsburgh, 1997; pp. 375-380.
13 Ryder, M. L. *J. Archaeol. Sci.* **1984**, *11*, 337-343.
14 Bengtsson, S. *Int..J.Nautical Archaeology.and Underwater Exploration* **1975**, *4*, 27-41.
15 Morris, K.;. Siefert, B. L. *J. Am. Inst. Conserv.* **1992**, *18*, 19-32.
16 Jenssen, V. In *Conservation of Marine Archaeological Objects*; Pearson, C. Ed.; Butterworths: London, 1987; pp. 122-163.
17 Peacock, E. E. *Int. Biodeterioration and Biodegradation* **1996**, *38*, pp. 49-59.
18 Peacock, E. E. *Int. Biodeterioration and Biodegradation* **1996**, *38*, pp. 35-47.
19 Florian, M-L. E. In *Conservation of Marine Archaeological Objects*; Pearson, C. Ed.; Butterworths: London, 1987; pp. 1-20.

20 Robson, R. M. In *Handbook of Fiber Chemistry*; Lewis, M.; Pearce, E. S. Eds.; Marcel Dekker: New York, 1998; pp. 416-460.
21 Lucas, F.; Shaw, J. T. B.; Smith, S. G. In *Advances in Protein Chemistry Vol XIII*; Anfinsen, C. B. Ed.; Academic: New York, 1958; pp. 106-242.
22 Light, A. *Proteins: structure and function*; Prentice-Hall: Englewood Cliffs, 1974.
23 Schulz, G. E.; Schirmer, R. H. *Principles of Protein Structure*; Springer-Verlag: New York, 1978.
24 Ishikawa, H.; Sofue, H.; Matsuzaki, K. *J.Text. Sci.Tech.Shinshu Univ.* **1960**, *10*, 176-178.
25 Shaw, J. T. B. *Biochem.J.* **1964**, *93*, 45-54.
26 Goldstein, J. I.; Newbury, D. E.; Echlin, P.; Joy, D. C.; Romig, A. D.; Lyman, C. E.; Fiori, C.; Lifshin, E. *Scanning electron microscopy and x-ray microanalysis*; Plenum: New York, 1990
27 Bhat, N. V.; Nadiger, G. S. *J. Polym. Sci.* **1980**, *25*, 921-932.
28 Freddi, G.; Romano, M.; Massafra, M. R.; Tsukada, M. *J.Polym.Sci., Polym.Phys Ed.* **1995**, *56*, 1537-1545.
29 Freddi, G.; Pessina, G.; Tsukada, M. *Int.J.Biol.Macromol.* **1999**, *24*, 251-263.
30 Fraser, R. D. B.; Suzuki, E. In *Physical Principles and Techniques of Protein Chemistry, Part B*; Leach, S. Ed.; Academic: New York, 1970; pp. 213-273.
31 Anonymous. Opus Spectroscopic Software Version 2.0 Reference Manual.. Bruker Instruments: Boston, 1995.
32 Kauppinen, J. K.; Moffatt, D. J.; Mantsch, H. H.; Cameron, D. G. *Appl.Spectrosc.* **1981**, *35* (3), 271-276.
33 Yang, C. Q. *Ind.Eng.Chem.Res.* **1992**, *31*, 617-621.
34 Becker, M. A.; Tuross, N. C. *Materials Res. Soc. Bull.* **1992**, *12* (1), 34-44.
35 Hirabayashi, K.; Yanagi, Y.; Kawakami, S.; Okuyama, K.; Hu, W. *J.Seric.Sci.Jpn.* 1987, **56** (*1*), **8**-22.
36 Becker, M. A.; Tuross, N. C. In *Silk Polymers*; Kaplan, D.; Farmer, B.; Viney, C. Eds.; American Chemical Society: Washington, D.C. 1994; pp. 252-269.
37 Nakanishi, M.; Koboyashi, K. *J.Soc.Text.Cell.Ind.Jpn.* **1954**, *10*, 128-131.
38 Sitch, D. A.; Smith, S. G. *J.Text.Inst.* **1957**, *48*, T341-T355.
39 Akune, S.; Koga, K. *Bull.Fac.Agr.Kagoshima Univ.* **1953**, *2*, 103-112.
40 Pérez-Rigueiro, J.; Viney, C.; Llorca, J.; Elices, M. *Polymer* **2000**, *41*, 8433-8439.
41 Tsukada, M.; Obo, M.; Kato, H.; Freddi, G.; Zanetti, F. *J.Appl.Polym.Sci.* **1996**, *60*, 1619-1627.
42 Jellinek, H. H. G. In *Physicochemical aspects of polymer surfaces, Vol. 1*; Mittal, K. L. Ed.; Plenum: New York, 1983; pp. 255-281.
43 Horne, R. A. *Marine chemistry*; Wiley-Interscience: New York, 1969.
44 Egerton, G. S. *Text.Res.J.* **1948**, *18*, 659-669.
45 Tímár-Balázsy, A.; Eastop, E. *Chemical principles of textile conservation*; Butterworth-Heinemann: Oxford, 1998.

## Chapter 10

# Electron Spin Resonance Studies To Explore the Thermal History of Archaeological Objects

**Robert G. Hayes[1] and Mark R. Schurr[2]**

**Departments of [1]Chemistry and Biochemistry and [2]Anthropology, University of Notre Dame, Notre Dame, IN 46556**

The heating of many organic materials produces species with unpaired electrons in the sample. These species may be studied by ESR. The characteristics of the ESR spectra depend on the temperaure to which the sample has been heated and on the time of heating, among other parameters. Thus, ESR may be used to refine the characterization of burned organic archaeological materials. We review work on charred cereal grains and its application to samples of charred maize recovered from six late prehistoric sites in the Midwest. By correlating the thermal histories of the prehistoric kernels with their archaeological contexts, we are able to control for some taphonomic biases in archaeobotanical samples. We also report ESR studies of modern bone samples heated under controlled conditions. They behave in the same way as carbonized plant remains at temperatures below about 300 °C, suggesting that ESR spectra can also be used to deduce the thermal histories of burnt bones from archaeological deposits.

It has been known for a long time that pyrolysis of organic materials produces paramagnetic materials *(1-3)*, and that the paramagnetism is very persistent. If the pyrolysis is carried out below 600 °C or so, the paramagnetism is believed to be due to free radicals that are produced in intermediate stages of the process by which heteroatoms are eliminated from the material and a carbon network is formed *(4 ,5)*. These free radicals are often very stable, and persist in the sample over hundreds or even thousands of years. The number and nature of these radicals can be investigated with electron spin resonance (ESR) spectra.

The ESR spectra of organic materials that have been pyrolysed often consist of a single rather narrow resonance *(6)*. The ESR parameters of the resonance depend on the temperature to which the sample had been heated and on the time of heating and, to a lesser extent, on other conditions of pyrolysis.

The availability of information about the thermal history of pyrolysed organic materials has several interesting potential archaeological applications. Charred, or apparently charred, plant remains, especially wood and cereal grains, are often found in archaeological contexts. Charred bone is also often found in interments. The ability to measure the temperature to which charred bone had been exposed is especially interesting, because bone collagen is often used in paleodiet studies and it is known that charring changes the isotope ratios of bone collagen *(7)*.

Hillman and co-workers have studied the ESR of charred cereal grains, primarily emmer wheat, and have shown that the g value of the resonance depends on the temperature of charring *(8, 9)*. More recently, we have studied the ESR of charred maize kernels *(10)*. Our studies complement the work of the Hillman group. We have used our results to discuss the thermal histories of charred maize kernels from six archaeological sites in the Midwest.

In this report, we review our work on maize, extend it somewhat, and attempt to provide some guidelines for the application of the work to archaeological samples.

We also, in a separate section of the report, make preliminary report of our work on the ESR spectra of samples of heated modern bone and the correlation of the ESR spectra with $^{13}C/^{12}C$ and $^{15}N/^{14}N$ ratios from the collagen of some of the samples. The isotope ratio measurements have allowed us to relate the ESR parameters, in particular the g value, to heating temperature, and has also allowed us to relate isotope ratio shifts to heating temperature or, more directly, to ESR g values.

## ESR and Charred Plant Remains

Hillman and co-workers *(8, 9)* studied the ESR spectra of charred cereal grains, especially emmer wheat. They found that the g factors of the spectra depended on the temperature of charring, but were not influenced by the time of exposure to the elevated temperature. Hillman and co-workers also found that the line width of the resonance, and its intensity, depended on the heating temperature, and could be used to support the g value in deducing the thermal history of a sample. They also found that the sample need not be freed of oxygen in order to be observed and that the g value of the resonance did not depend on the amount of oxygen adsorbed or absorbed by the sample. Our work confirmed all these results.

The primary quantitative result of the study by Hillman et al. *(9)* was a calibration curve of g values with temperature, claimed to be valid, independently of heating conditions (ranging from heating in air to heating in vacuum) and, by implication, applicable to all plant materials. Hillman and co-workers were able to obtain data over a temperature range of 20 °C to 600 °C. They found a monotonic decrease in the g value of the resonance, from 2.0043 at room temperature to 2.0024 at 600°C *(Cit. 9, Fig. 3)*.

We prepared, and measured triplicate sets of samples. We examined three conditions of charring. These were charring in air, charring in sand and charring in evacuated sealed tubes. We also examined two times of exposure to the elevated temperature, five minutes and two hours. There was little difference between the parameters of samples charred for five minutes and those charred for two hours, except at low temperatures. We attribute this to failure of the samples to come to temperature in five minutes at low temperatures. The samples charred in sand and the samples charred in air had very similar g values, and the intensities were, on the whole, quite similar.

Figures 1 and 2 show our results for samples that were charred for two hours, in air and in evacuated tubes. We show mean values and standard deviations of the parameters. The standard deviations are shown as error bars in the figures.

Our results, clearly, parallel the earlier results, but there are some significant differences. Also, we performed some ancillary experiments to examine certain aspects of the results, especially aspects that might have relevance to archaeological applications of the work. We discuss these matters here.

Firstly, we were never able to take a sample above 400°C because the samples burned above this temperature. Thus, we never obtained a sample with a g value as low as 2.0024 and, in fact, the g values we observed at the highest temperature we could achieve, 400°C, were considerably higher than the value that Hillman and co-workers report for 400°C (2.0026).

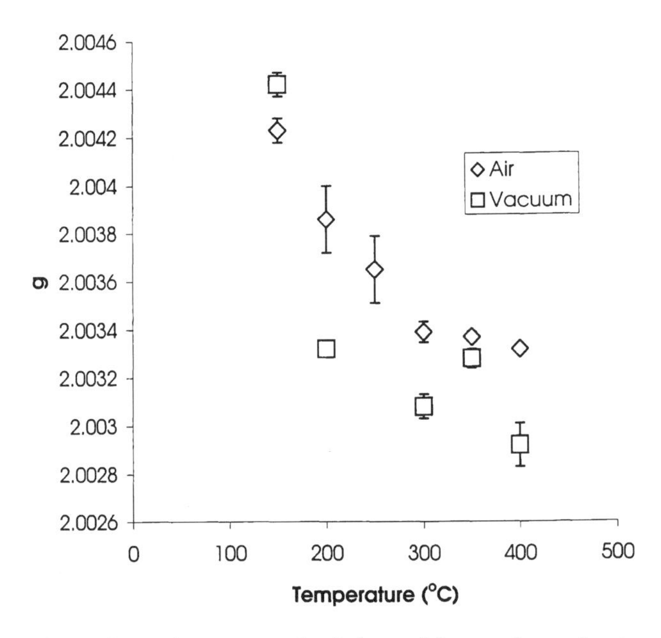


*Figure 1. g values of maize standards heated for two hours in air and in evacuated tubes. Standard deviations are shown as error bars.*

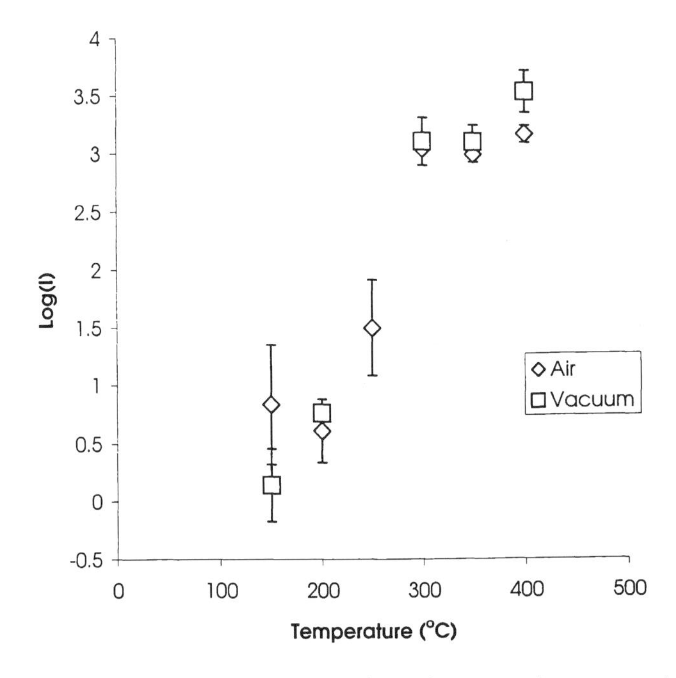


*Figure 2. Intensities of maize standards heated in air and in evacuted tubes. Standard deviations are shown as error bars.*

The discrepancy between our results and the previous results appears to be due, at least, in part, to different ways of heating the samples. Hillman and co-workers claimed that the ESR parameters of the samples were insensitive to the details of the heating process. The results that they reported came, in fact, from samples that were heated in ESR tubes. The samples were, thus, in contact with the gaseous products of pyrolysis during the heating. These shielded the samples from the atmosphere. In addition, it is known that the course of charring is different if the sample is kept in contact with pyrolysis products (11 ) .

We tested the hypothesis that charring in an ESR tube produces a different g value from charring under conditions that make air more readily available to the sample, by charring maize kernels simultaneously in sand and in ESR tubes at 300 °C for 2 hours. The samples charred in sand had g = 2.0034. Those charred in ESR tubes had g = 2.0031.

Our data and the earlier data on standards impose some limitations on the application of the results to archaeological samples. Firstly, it is clear from the scatter in our standard data, and the stated uncertainty in g of ±0.0002 in the earlier data, that an ESR spectrum from a single ancient sample cannot yield a well-defined charring temperature, but only a temperature in a range of, perhaps, ± 50 °C.

Secondly, it is clear that the ESR data from a charred sample depend somewhat on the conditions of charring, as we have discussed. Comparison of data from ancient samples with data from standards which were charred under known conditions may allow one to suggest the conditions under which the ancient samples were heated. We apply this idea in our analysis of ancient maize samples, presented in the next section.

Inferences about ancient charring conditions based on the ESR spectra of modern samples charred under contolled conditions will only be reliable if ESR parameters remain consistent over long periods of time. There is no obvious way to experimentally assess the effect of relatively long times on ESR parameters for carbonaceous chars. Since ancient samples are, for the greater part, exposed to the oxygen in air and groundwater, one might imagine that the g value of a sample would rise with time, as oxygen reacts with the sample. As a first attempt to examine this question experimentally, we took duplicate samples of pieces of a maize kernel, that had been charred for 4 hours at 250°C in sand, and held them at 120°C for two weeks. The g values rose by 0.00019 and 0.00015, and the intensities fell to 67% and 77.8% of the original values, suggesting that small increases in g and decreases in intensity should be expected in prehistoric samples compared to modern controls heated under the same conditions.

## An Application to Prehistoric Maize Samples

The Fort Ancient culture *(12)* represents the remains of a group of prehistoric tribal societies that inhabited the central Ohio River Valley of eastern North America after A.D. 1000. At the same time, Middle Mississippian chiefdoms were present in the lower Ohio Valley. Both types of societies were maize agriculturalists, and archaeologists have long been interested in knowing what role maize played in the subsistence systems of these two different types of prehistoric societies.

Archaeobotanical data *(13, 14)* have been cited as evidence that Fort Ancient populations were more heavily dependent upon maize than the Middle Mississippians were. Higher ubiquities and densities of maize remains were found in flotation-processed samples from Fort Ancient sites compared to samples from Middle Mississippian ones. In this region, stable carbon-isotope ratios (expressed as $\delta^{13}C$ values in units of per mil [‰]) are positively correlated with maize consumption *(15)*. Stable isotope ratios obtained for Middle Mississippian sites in the lower Ohio Valley *(16-18)* average between -9.5‰ and –8.7‰. These averages are more positive than those reported for most Fort Ancient sites *(17-21)*, all of which have produced average values ranging between -9.4‰ and -11.8‰, indicating that Fort Ancient diets of the middle Ohio Valley contained less maize on the average than did Middle Mississippian diets of the lower Ohio Valley.

The stable carbon-isotope data are in sharp contrast to those based on archaeobotanical data in this region. One possible explanation for the differences in the inferred relative rates of maize consumption drawn from the isotopic and archaeobotanical data is the fact that the two approaches measure different things. While isotope ratios provide a measure of individual maize consumption, archaeobotanical data reflect maize remains that were not consumed, and are the result of prehistoric food processing and storage techniques, combined with the taphonomic processes and archaeological recovery methods that produced the samples of carbonized maize remains available to the archaeobotanist. ESR measurements can be used to help understand the taphonomic processes.

In order to evaluate the effects of carbonization temperature on the relative abundance of archaeobotanical maize at prehistoric sites, ESR spectra were obtained from 70 prehistoric maize kernels and kernel fragments from six different prehistoric sites representing variants of the Fort Ancient and Middle Mississippian cultures in the Ohio Valley of North America between AD 1000 and 1600 *(10)*.

The maize remains had previously been recovered using two different sets of techniques. Excavations at four sites were conducted before the invention of flotation recovery techniques and botanical collections consist mainly of large, well-preserved maize kernels. All the samples used here apparently came from pit features. Collections from two Oliver phase sites that represent a regional variant of Fort Ancient in southern Indiana (Cox's Woods and Clampitt) were recently taken using flotation recovery *(22, 23)*. The maize remains from these sites consist of very small kernel fragments from pit feature fill and midden contexts.

The g values and spin intensities of specimens from each context showed substantial variation, revealing that a range of carbonization conditions were experienced by kernels even when they were depositied together in a single feature. Figures 3 and 4 shows the data from ancient samples, in the form of plots which place each sample in the g value/intensity plane. The data are superimposed on plots of the standards that were prepared in evacuated, sealed tubes. Midden samples had all been recovered by flotation. The pit samples that were recovered by flotation are distinguished in Figure 4 from those that were recovered without use of flotation.

The midden samples all have similar ESR parameters, and all seem to have been charred at a fairly high temperature, under conditions that excluded oxygen and kept the kernels in contact with pyrolysis gases. The ESR parameters cluster around those that correspond to a charring temperaure near 300 °C.

The ESR parameters of the pit samples are, clearly, much more varied. On the whole, the flotation samples from pits also fit into the data from standards that were prepared in evacuated tubes. This is to be expected if they were produced when fires were built in storage pits to fumigate them. The ESR parameters range over values that correspond to charring temperaures between 250 °C and 350 °C.

The majority of non-flotation samples from pits do not fit into the data from pyrolysis in evacuated tubes very well, although the samples that have lower g values do fit well. The others, with g values above 2.0035, fit better into the data from standards that were charred in air.

The g values of samples from pit features are generally higher than those of samples from midden contexts, while the spin intensities of pit feature samples are generally lower. This indicates that the small kernel fragments from midden contexts were heated to higher temperatures for longer periods of time than were the relatively intact kernels from pits. Because of their thermal history, the original mass of kernels in the midden was substantially reduced by the carbonization process and the fragments recovered by flotation will under-represent the amount of maize that was originally present.

The Middle Mississippian preference for above ground storage (corn cribs) contrasts strongly with the Fort Ancient preference for storage in subsurface storage pits *(13)*. Thus, at Middle Mississippian sites, maize remains will be almost exclusively or heavily biased towards midden samples. This was the case for the archaeobotanical data originally cited for Middle Mississippian sites *(13)*,

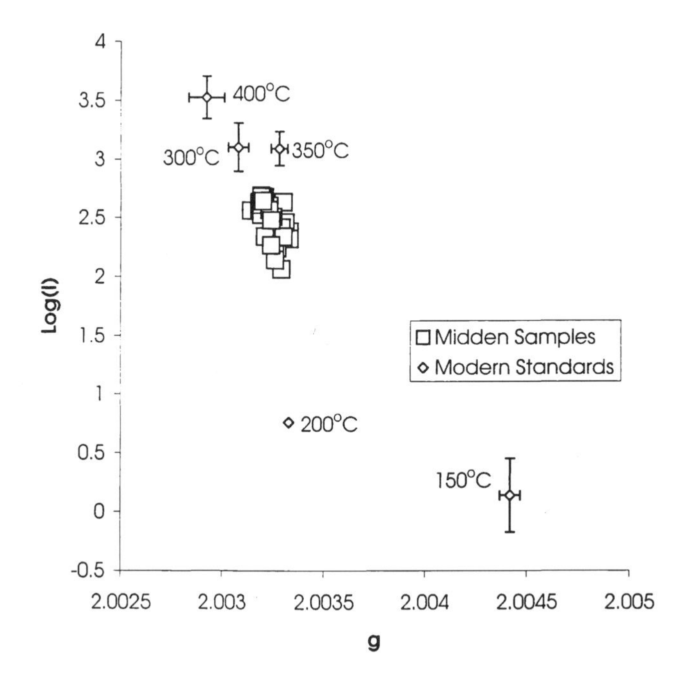


*Figure 3. Data from ancient maize samples recovered from middens, plotted on a g value / intensity plane. Data from heating in evacuated sealed tubes are included for reference.*

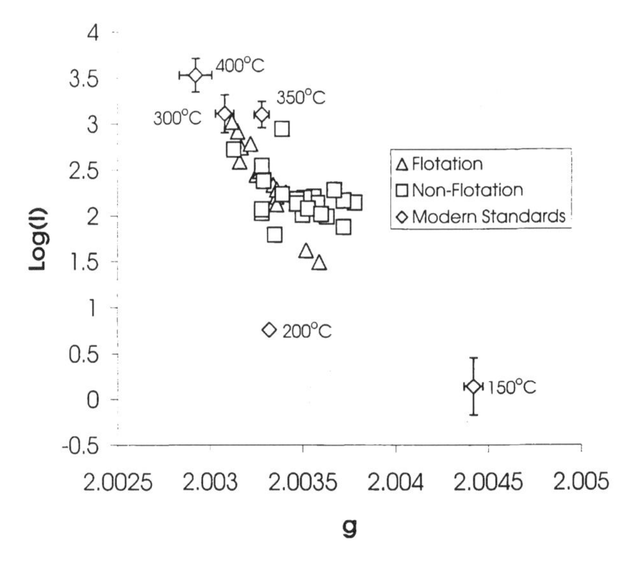


*Figure 4. Data from ancient maize samples recovered from pits, plotted on a g value / intensity plane. Data from heating in evacuated sealed tubes are included for reference.*

where Middle Mississippian maize samples came only from midden contexts. As the ESR spectra show, specimens from such contexts were usually carbonized at relatively high temperatures, and will therefore significantly under-represent the amount of maize originally present. The differing conclusions about relative maize consumption drawn from archaeobotanical remains and stable carbon-isotope ratios arose from an imperfect understanding of the taphonomic processes that produced the archaeobotanical remains. While it may seem obvious that samples can only be compared if they come from similar contexts, the ESR results show that, in additon to coming from similar contexts, comparative samples should have also experienced similar thermal histories. This example illustrates that ESR data can be used to understand how botanical remains enter the archaeological record so that the effects of taphonomic processes can be more accurately accounted for.

## The ESR Spectra of Heated Bones

Some time ago, Sales and co-workers observed the ESR spectra of ancient bones before and after heating *(24)*. All the ancient bone samples that they studied were at least 7500 years old, and most were older than 300,000 years. In these, aside from resonances that were due to transition metal ions in the bone, the spectra were dominated by the defect signal that is produced by radiation damage to the mineral component of the bone, and which is the signal that is used to date bone and tooth material. This signal is not observable in bone or tooth that is less than about 30,000 years old *(24)*. The temperature dependence of the spectra from the samples that Sales and co-workers observed presumably represents the temperature dependence of the spectrum of this defect.

More recently, Michel and co-workers have studied charred modern bone by ESR, IR and x-ray diffraction, and have also studied Middle Pleistocene fossil bone from Lazaret Cave in France *(25)*. They found that the IR feature that they took to be the signature of the organic component of the bone disappears from the spectra of bone that had been heated above about 400 °C. They also found that bone developed a characteristic ESR spectrum upon heating to temperatures between 200 °C and 600 °C, and that this spectrum was replaced by a different, much weaker, spectrum in samples that were heated above 600 °C.

### ESR Measurements on Bones Heated Under Controlled Conditions

Samples of modern bone were taken from a beef rib that had been dried at 40 °C for an extended period of time. Samples ranged in size from 50 mg to 200 mg. The samples were heated in a thermostatically-controlled muffle furnace for periods of time that ranged from 0.5 hour to 4 hours. Weight loss from heating was obtained for each sample. The carbon and nitrogen contents of the burned bones were measured using a Carlo-Erba elemental analyzer. Collagen and acid-insoluble charred organic materials were then extracted by slow demineralization in weak acid using procedures commonly used for isotopic studies of prehistoric bones *(26)*. Stable carbon- and nitrogen-isotope ratios of the extracted organic materials were determined with a Finnegan Delta Plus mass spectrometer equipped with a Carlo-Erba elemental analyzer and a Conflo II interface housed at the Center for Environmental Science and Technology, University of Notre Dame.

ESR spectra were measured on a Bruker EPR spectrometer at room temperature. Spectra were obtained at the lowest power and the smallest modulation field that was consistent with good sensitivity. Standard samples that had been heated to temperatures above 200 °C could be examined using a microwave power of 0.01 mW and a modulation amplitude of 0.3 gauss peak-to-peak. Samples that had never been heated, or that had been heated to temperatures below 200 °C required higher power and, sometimes, a greater modulation amplitude. The spectrum of Figure 5, for instance was obtained at a power of 1.0 mW and a modulation ampitude of 0.10 gauss peak-to-peak. In no case was a power greater than 1.0 mW or a modulation amplitude greater than 1.0 gauss peak-to-peak used. The microwave frequency of the spectrum is available from a frequency counter that is part of the spectrometer. The magnetic field at the center of the spectrum was obtained from the output of the calibrated Hall probe that is used to control the magnetic field. The magnetic field indicated by the Hall probe was checked periodically against a nuclear resonance gaussmeter.

Spectroscopic splitting factors were calculated from the microwave frequency and the magnetic field at the center of the spectrum. The spectra were doubly integrated, using the spectrometer software, to give experimental intensities. These were normalized to standard sample mass and spectrometer parameters to obtain quantities that are, approximately, proportional to the density of unpaired electrons in the sample. We call these quantities intensities (I).

The ESR spectra from bone samples fall into three types, which are correlated with the thermal history of the samples. Bone standard samples that have not been heated above the temperature needed to dry them (40 °C), and prehistoric bone samples that appear never to have been heated, display a weak ESR signal. The dried standards have a signal that seems to correlate with the signals of heated bone standards, but is very weak. The (apparently) unheated prehistoric bone samples have a signal that has $g = 2.0040$ at the peak of the spectrum. This spectrum appears to be that of a randomly oriented species with three distinct g values. It is hard to fit the spectrum to a simulated spectrum because it is weak, and is underlain by a broad, stronger spectrum, presumably due to Fe(III) ions in the sample. Figure 5 shows the spectrum from a sample of human prehistoric bone that is about 5,000 years old (a rib fragment of Burial 190, a 25 – 35 year old male burial from the Indian Knoll site [15 OH 2] *[27]*), and an attempt to fit the spectrum. The simulated spectrum has a Lorentzian line shape, 3.0 gauss wide, independent of angle, and $g_1 = 2.0062$, $g_2 = 2.0039$, $g_3 = 2.0015$. It provides a reasonable fit to most of the experimental spectrum, but the experimental spectrum seems to have intensity in the wings that is unaccounted for by the simulated spectrum.

Bone standards that have been heated to temperatures between 150 °C and about 350 °C, and prehistoric bone samples that have, presumably, experienced temperatures in the same range (based on their similar coloration), display a nearly symmetric ESR spectrum. The g value of the spectrum from the standards depends on the heating temperature, as shown in Figure 6.

Above a heating temperature of about 350 °C, the spectra of the standard samples become independent of heating temperature. The spectrum is not symmetric. It can be fit to the expected spectrum of a randomly oriented species with three distinct g values. An example of such a spectrum, and an attempt to fit the spectrum, appears in Figure 7. The simulated spectrum that was used to fit the spectrum has g values $g_1 = 2.0053$, $g_2 = 2.0039$, and $g_3 = 2.0020$. The spectrum was assumed to have a Lorentzian line with a width of 2.00 gauss, independent of angle. As one sees, the simulated spectrum reproduces most of the experimental spectrum well. Once again, however, the intensities in the wings of the spectrum are not reproduced by the simulated spectrum. The g value at the peak of the integrated spectrum is 2.0043. In the temperature range near 350 °C, the observed spectrum appears to be a superposition of a symmetric spectrum and the high-temperature spectrum.

Perhaps the most interesting of our results is the relationship between heating temperature and the g value of the ESR spectrum shown in Figure 6, a behavior that is similar to the behavior of plant remains. We have verified that the spectra we observe come from the organic part of the bone by observing the ESR spectrum from the collagenous material of some of the samples that had been demineralized for isotopic analysis. These spectra had intensities that were 30% to 50% of those of the bone samples from which the collagen came. This is within the range of uncertainty of intensities, as we define them, due to variations in spectrometer conditions.

The results shown in Figure 6 permit one to use the ESR signal as a thermometer to assess the maximum temperature to which a bone sample has been heated, in the range from about 100 °C to some 300 °C.

Bone that has been heated above about 350 °C reveals this fact clearly in the replacement of the symmetric spectrum of charred collagen by the spectrum of Figure 7. This spectrum is most easily recognized from its g value, near 2.0043,

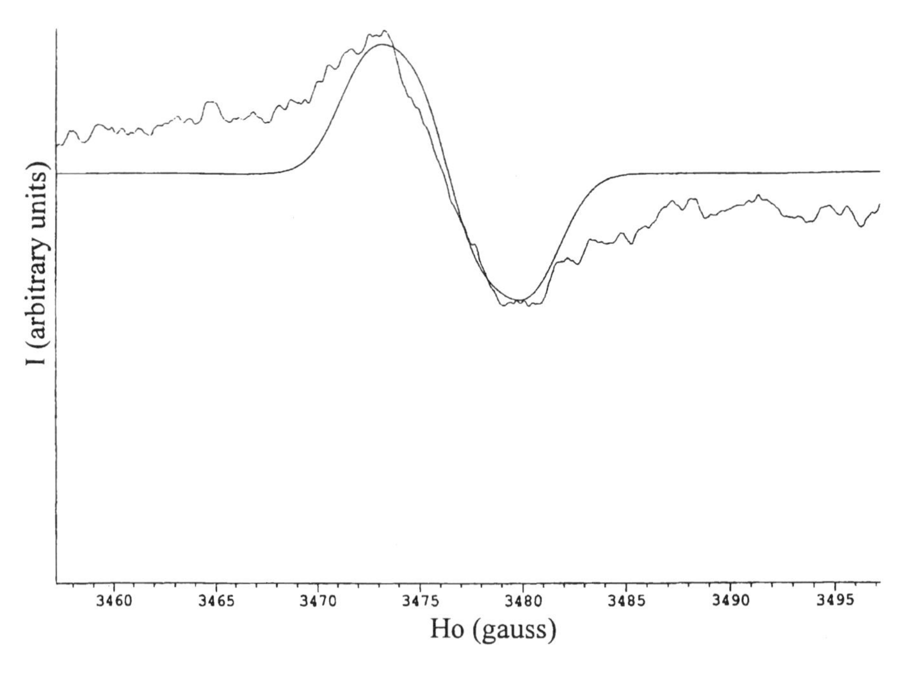


*Figure 5. ESR spectrum of unheated ancient bone samples and a simulation of the spectrum. See text for details*

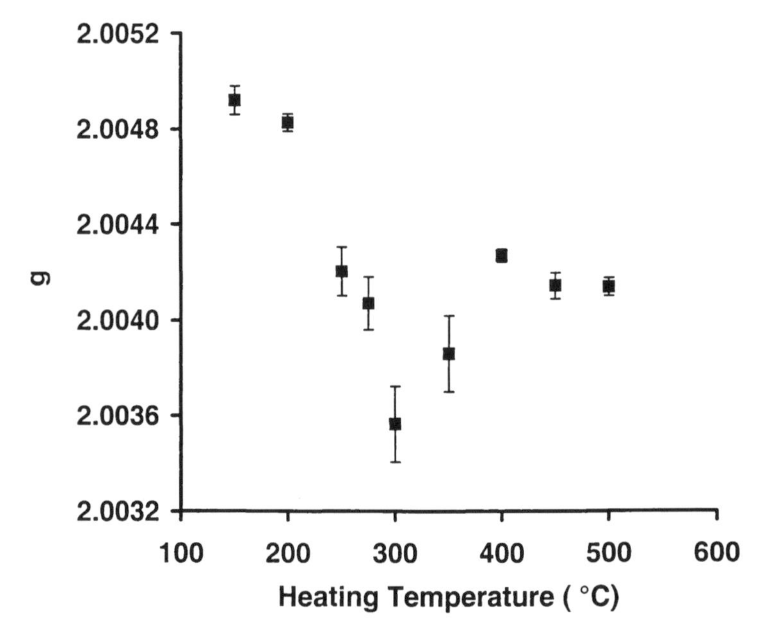


*Figure 6. Changes in the g values with temperature for modern bone samples.*

and the fact that it is unsymmetrical. The asymmetry can be quantitated by measuring the height of the downfield and upfield maxima in the ESR spectrum and taking the ratio of their absolute values. This is shown, on a spectrum from a standard heated to 400 °C, on Figure 8. The ratio for the spectra from charred collagen, i.e., bone samples heated to a temperature less than 300 °C, is near 1.00. That from the spectrum observed above 350 °C is near 1.30. Bone that has been heated above 350 °C can easily be distinguished from unheated bone by its chalky (calcined) appearance. Failing this, the ESR spectra, shown in Figures 5 and 7, are distinct.

The sources of the spectra are of interest. As we have mentioned, the symmetric spectrum that is observed in samples that have been heated between 100 °C and 350 °C can be assigned to charring of the organic material in the bone. It is believed that this material has patches that are mostly carbon networks, embedded in less carbonized material. The patches contain one or more unpaired electrons. As heating progresses, the patches eventually coalesce, with pairing of the electrons, and concomitant reduction of the ESR signal.

The spectrum observed above 350 °C, fitted in Figure 7, resembles, in a general way, the spectra of species that are created by irradiation of carbonate-containing hydroxyapatite *(28)*. Callens and co-workers have found that irradiated carbonate-containing hydroxyapatites display ESR spectra that are the sum of contributions from several species *(28)*. In particular, they display spectra that have been assigned to $CO_2^-$ in three environments, and also to $CO_3^{3-}$ in three environments. The spectra we observe do not appear to have contributions from $CO_2^-$. This radical is characterized by one small g value, near 1.997, that produces a high-field shoulder in the spectra. We do not observe such a feature. As one sees from Figure 5, the difference spectrum between the spectrum we observe and the simulation, is broad and has no feature at g = 1.997. The spectra of the $CO_3^{3-}$ species are more similar to the spectra we observe. In particular, a spectrum that Callens and co-workers assign to a phosphate site has $g_1 = 2.0047$, $g_2 = 2.0035$, $g_3 = 2.0018$. Neither this spectrum, nor any other spectrum that has been assigned to $CO_3^{3-}$, has a g value as high as 2.0053. It seems likely that the spectra we observe are due to $CO_3^{3-}$, however, perhaps the $CO_3^{3-}$ ions are located on several sites that include hitherto unobserved sites on the surfaces of apatite crystals.

The spectrum that is observed in ancient bone that has not been heated does not resemble the spectrum of any assigned species of which we are aware. Such spectra have sometimes been assigned to an undefined radical in the organic part of the bone *(24)*, or to peroxidation of unsaturated fats in the bone *(29)*. The characteristics of the spectra that are discussed in these two citations have higher g values than those we observe, however. (g = 2.0046 *[24]*, g = 2.0045 *[29]*).

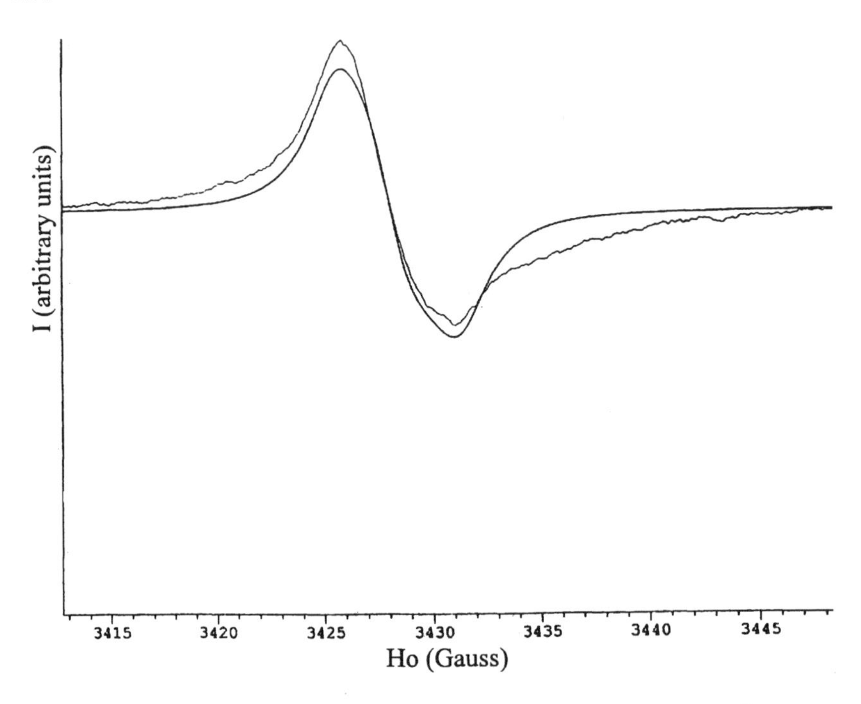


*Figure 7 ESR spectrum of a sample of modern bone, heated to 400 °C, and a simulation of the spectrum. See text for details.*

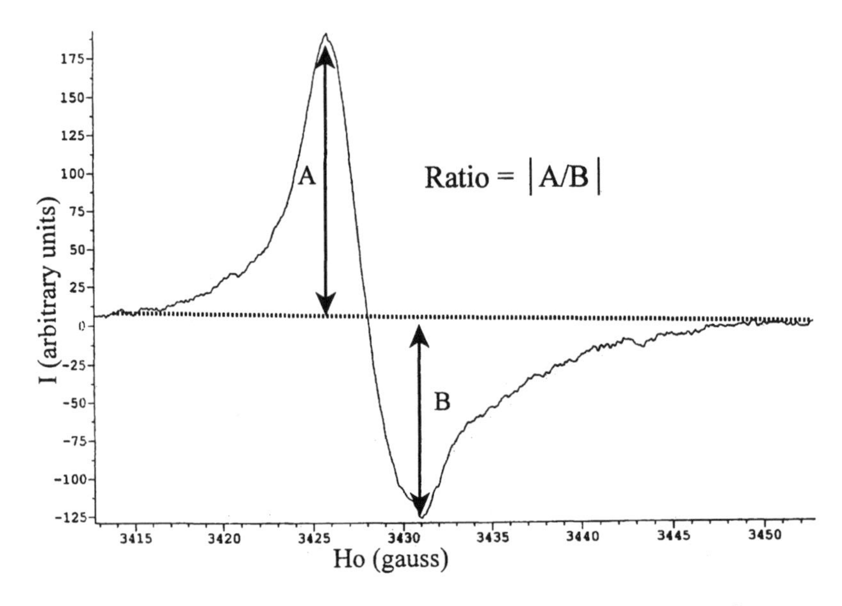


*Figure 8. ESR spectrum of a sample of modern bone, heated to 400 ° showing how the asymmetry of the spectrum is measured.*

## Covariation in ESR Spectra and Stable Isotope Ratios with Heating Temperature

Bone goes through a well-defined sequence of chemical change with increasing temperature. As summarized by McCutcheon *(30)* temperatures between 20 °C and 300 - 350 °C lead to loss of unbound water and initial carbonization of the organic phase, while temperatures above 350 °C result in complete combustion of the organic phase. Data published by DeNiro et al. *(7)* show changes in stable carbon- and nitrogen-isotope ratios with heating temperature. While their data are limited and contain much scatter, the isotopic changes they observed were relatively regular for conditions where the collagen content remained above about 2%. The 2% threshold is significant because this limit is often cited as the minimum level of collagen preservation required for reliable isotope ratios *(31)*. DeNiro et al., observed decreased $\delta^{13}C$ values and increased $\delta^{15}N$ values with higher temperatures or longer heating times in heated bones retaining more than 2% of the original organic material *(7)*. Thus, if one knew the heating times or temperatures of the specimens, it might be possible to reconstruct the isotopic composition of the original sample by correcting for the isotopic effects of thermal alteration.

As expected from earlier studies of burned bone *(30)*, the carbon and nitrogen contents (expressed as %C and %N) and the extraction yield declined sharply at temperatures above 350 °C (Figure 9). Stable carbon- and nitrogen-isotope ratios remained unaltered at 200°C, and then respectively declined and increased between 250°C and 400°C (little or no organic material could be extracted from samples heated at 450°C). These results are consistent with those reported by DeNiro et al. *(7)*. At temperatures below 350°C, $\delta^{13}C$ values are strongly correlated with g values in the demineralized extracts (Figure 10). We found similar but less clear trends relating spin intensities and peak widths to heating times and temperatures, but these factors varied in a more complex way, so additional control samples will be needed to quantify these relationships. As was the case for carbonized plant remains, spin intensities may be strongly related to heating time, so that it may be possible to use g values and spin intensities to determine both heating times and temperatures.

Based on these results, it is clear that g values can be used to reconstruct the thermal history of burned bones, and that correction factors can be developed to "back correct" for the effects of thermal alteration on stable carbon- and nitrogen-isotope ratios. Further experiments are needed on charred and uncharred bones from prehistoric sites to test the accuracy of "back corrections" in prehistoric cremated bones.

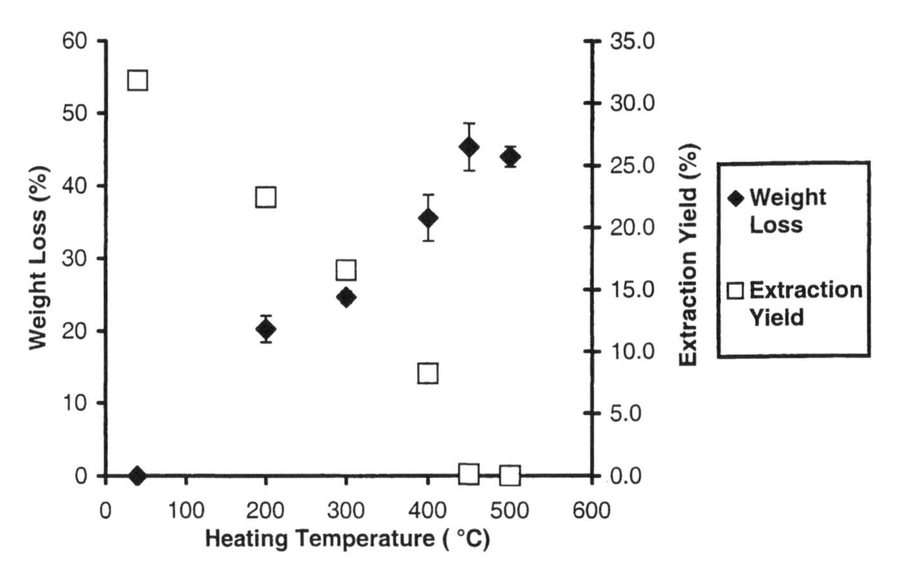


*Figure 9. Weight loss and extraction yields after heating show reductions of the organic phase with increased temperatures.*

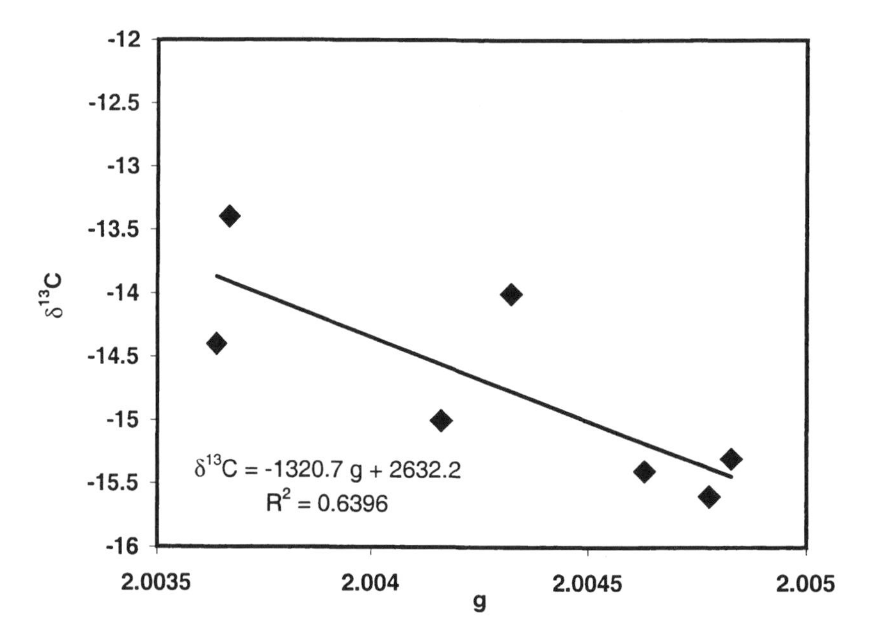


*Figure 10. The $\delta^{13}C$ values of extracts are correlated with g values at temperatures below 350 °C.*

# CONCLUSIONS

ESR spectra reveal significant information about the thermal history of carbonized plant remains and bones. In particular, they can be used to determine the maximum heating temperature of carbonized plant remains, and provide some information about the heating duration. An ESR spectrum also shows whether a bone has been heated or not and, in the temperature range up to 300 °C or so, it is able to reveal the temperature to which the bone has been heated. This information can be used to study the taphonomy of burned bones and may be useful for expanding paleodietary studies to thermally altered bones.

## Acknowledgements

We are very grateful to Charles Kulpa and Dennis Birdsell for the support of the stable isotope facilities at the Center for Environmental Science and Technology, University of Notre Dame. The manuscript also benefited greatly from the comments of an anonymous reviewer and the efforts of Kathryn Jakes, who promptly forwarded comments to us while tolerating our last minute approach to deadlines.

## References

1. Ingram, D. J. E.; Bennett, J. E. *Philos. Mag.* **1954**, *45*, 545-547.
2. Uebersfeld, J.; Etienne, A.; Combrisson, J. *Nature* **1954**,174, 614.
3. Winslow, F. H.; Baker, W. O.; Yager, W. A. *J. Am Chem. Soc.* **1955**, *77*, 4751-4756.
4. Singer, L. S.; Lewis, I. C. *Applied Spectrosc.* **1982**, *36*, 52-57.
5. Tang, M. M.; Bacon, R. *Carbon* **1964**, *2*, 211-220.
6. Mrozowski, S. *Carbon* **1988**, *26*, 531-541.
7. DeNiro, M. J.; Schoeninger, M. J.; Hastorf, C. A. *J. Archaeol. Sci.* **1985**, *12*, 1-7.
8. Hillman, G. C.; Robins, G V.; Oduwole, A. D.; Sales, K. D.; McNeil, D. A. C. *Science* **1983**, *222*, 1235-1236.
8. Hillman, G. C.; Robins, G. V.; Oduwole, A. D.; Sales, K. D.; McNeil, D. A. C. *J. Archaeol. Sci.* **1985**, *12*, 49-58.
10. Schurr, M. R.; Hayes, R.; Bush, L. L. *Archaeometry* **2001**, *43*, 407-419.
11. Mrozowski, S. *Carbon* **1981**, *19*, 365 – 370.

12. Griffin, J. B. *The Fort Ancient Aspect: Its Cultural and Chronological Position in Mississippi Valley Archaeology*; University of Michigan Press: Ann Arbor, MI, 1943.
13. Rossen, J.; Edging, R. In *Current Archaeological Research in Kentucky;* Hockensmith, C.D.; Carstens, K. C.; Stout, C.; Rivers, S. J., Eds.; Kentucky Heritage Council: Frankfort, KY, 1987; Vol. 1, pp 225-234.
14. Rossen, J. In *Fort Ancient Cultural Dynamics in the Middle Ohio Valley;* Henderson, A. G., Ed.; Prehistory Press, Madison, WI, 1992; pp 189-208.
15. Vogel, J. C.; van der Merwe, N. J. *Am. Antiq.* **1977**, *42*, 238-242.
16. Schurr, M. R. *Am. Antiq.* **1992**, *57*, 300-320.
17. Schurr, M. R.; Schoeninger, M. J. *J. Anthropo. Archaeol.* **1996**, *14*, 315-339.
18. Schurr, M.R. In *Current Archaeological Research in Kentucky;* Hockensmith, C. D.; Carstens, K. C.; Stout, C.; Rivers, S. J., Eds.; Kentucky Heritage Council: Frankfort, KY, 1998; Vol. 5, pp. 233-257.
19. Broida, M. O. In *Late Prehistoric Research in Kentucky*; Pollack, D.; Holkensmith, C.; Sanders, T., Eds; Kentucky Heritage Council: Frankfort, KY, 1984; pp. 68-83.
20. Conard, A. R. In *A History of 17 Years of Excavation and Reconstruction - A Chronicle of 12th Century Human Values and the Built Environment*; Heilman, J.; Lileas, M.; Turnbow, C., Eds; Dayton Museum of Natural History: Dayton, OH, 1988; pp. 112-156.
21. van der Merwe, N. J.; Vogel, J. C.; Nature **1978**, *276*, 815-816.
22. McCullough R. G.; Wright, T. M.; *An Archaeological Investigation of Late Prehistoric Subsistence- Settlement Diversity in Central Indiana*, Glenn A. Black Laboratory of Archaeology Research Reports, No. 18. Indiana University: Bloomington, IN, 1997.
23. Redmond, B. G. *The Archaeology of the Clampitt Site (12 Lr 329), an Oliver Phase Village in Lawrence County, Indiana*, Glenn A. Black Laboratory of Archaeology Research Reports, No. 16. Indiana University: Bloomington, IN, 1994.
24. Sales, K. D.; Oduwole, A. D.; Robins, G. V.; Olsen, S. *Nucl. Tracks* **1985**, *10*, 845-851.
25. Michel, V.; Falgueres, C.; Dolo, J.-M. *Radiat. Meas.* **1998**, *29*, 95-103.
26. Moore, Murray and Schoeninger, *J. Archaeol. Sci.*, **1989**, *16*, 443-446.
27. Snow, C. E. *Indian Knoll Skeletons of Site Oh2, Ohio County, Kentucky.* University of Kentucky Reports in Anthropology and Archaeology No. 4. University of Kentucky, Lexington, 1948.
23. Amira, S.; Vanhaelwyn, G.; Callens, F. *Phys. Chem. Chem. Phys.* **2001**, *3*, 1724-28
24. Ikeya, M. *New Applications of Electron spin resonance: Dating, Dosimetry and Microscopy*; World Scientific: Singapore, 1993; pp. 247-252.
25. McCutcheon, P. T. In *Deciphering a Shell Midden*; Stein, J. K., Ed.; Academic Press: New York, 1992; pp. 347-370.
26. Ambrose, S. H. *J. Archaeol. Sci.* **1990**, *17*, 431-451.

## Chapter 11

# Geochemical Analysis of Obsidian and the Reconstruction of Trade Mechanisms in the Early Neolithic Period of the Western Mediterranean

**Robert H. Tykot**

**Laboratory for Archaeological Science, Department of Anthropology, 4202 East Fowler Avenue, SOC 107, University of South Florida, Tampa, FL 33620**

Geochemical fingerprinting of obsidian sources was first applied in the Mediterranean region nearly four decades ago. Since then, a number of analytical methods have proven successful in distinguishing the western Mediterranean island sources of Lipari, Palmarola, Pantelleria, and Sardinia. The existence of multiple flows in the Monte Arci region of Sardinia and on some of the other islands, however, has enabled the study of specific patterns of source exploitation and the trade mechanisms which resulted in the distribution of obsidian hundreds of kilometers away during the neolithic period (ca. 6000-3000 BC). Results are presented here from the chemical analysis of significant numbers of artifacts from several Early Neolithic sites in Sardinia, Corsica, and northern Italy as part of the largest and most comprehensive study of obsidian sources and trade in this region. The patterns of source exploitation revealed by this study specifically support a down-the-line model of obsidian trade during this period.

Obsidian tools found at prehistoric sites in the Mediterranean are evidence of a complex series of activities including procurement and transport of the raw material from island sources, production and distribution of cores and finished tools, use and eventual disposal. The identification of the geological source establishes the beginning of this *chaîne opératoire*, while the analysis of geographic and chronological patterns of obsidian source exploitation complements lithic reduction and use-wear studies in elucidating the intervening behavioral links. Obsidian is perhaps the most visible indicator of interactions during the Neolithic period (ca. 6000-3000 BC), and as such is important not only to our understanding of long-distance exchange networks and craft specialization, but also of the earliest settlement of the Mediterranean islands and the transition from hunting and gathering to an agricultural way of life.

## The Early Neolithic in the Western Mediterranean

The Early Neolithic in the western Mediterranean is defined by the first appearance of ceramics along with domesticated plants and animals. This transition went hand in hand with an accelerated involvement of Sardinia and Corsica in Mediterranean interrelations (*1*). Several dozen Early Neolithic sites have been identified on these islands, including caves and rockshelters concentrated in the less mountainous parts of the islands or near the coasts, but also including open-air sites. Some are located well in the interior of the islands, away from fluvial systems. Radiocarbon dates place the Cardial phase of the Early Neolithic in Sardinia and Corsica at 5700-5300 cal BC (*2*).

Early Neolithic pottery decorated with impressions of *Cardium edule* shells is found not only in Sardinia and Corsica, but also in other areas of the western Mediterranean. At Filiestru Cave in Sardinia, for example, Cardial impressed bowls, plates, and jars comprise 7% of the ceramic assemblage (*3*). Provenance studies of these wares indicates the existence of intra-regional, multi-directional interaction networks, with vessels found up to 50-70 km from their production area (*4*). Overall, we may interpret the Cardial phenomenon as suggesting a common cultural base over much of the western Mediterranean, with broad inter-group interaction evidenced not only by the ceramics (*5, 6, 7*), but also by inter-regional movement of ground and chipped stone artifacts including obsidian (Figure 1).

Lithic assemblages are typically composed of geometric microliths, including rectangles, trapezes, lunates, triangles, and larger implements including scrapers, burins, and transverse tranchet arrowheads (*8, 9, 10*). These tools were fashioned from flint, quartz, rhyolite, and above all obsidian. In Corsica, obsidian is rare or non-existent in the earliest Cardial I phase, although the flint from which most tools were made was imported from the Perfugas area in Sardinia (*11, 12*). In Sardinia,

*Figure 1. Obsidian sources in the Central Mediterranean and archaeological sites with more than 10 analyzed obsidian artifacts.*

obsidian is found at all Early Neolithic sites, and accounts for 17% of the Cardial I lithic assemblage at Filiestru (*3*). In Cardial II, obsidian becomes abundant at Corsican sites, although obsidian cores are small and rare, and arrowheads are infrequently made of obsidian.

## Geological Sources of Obsidian

The existence of obsidian sources on the Italian islands of Sardinia, Lipari, Palmarola and Pantelleria has been well documented, and early provenance studies identified them as the source of virtually all obsidian artifacts found at archaeological sites in the western Mediterranean (*13, 14*). Resolution has become one of the most important considerations in the characterization of western Mediterranean obsidian sources, since obsidian may be found in more than one locality on each island (*15*). While the geological age of obsidian formation on each island is sufficiently different for fission-track dating to have been widely used as a provenance technique (*16*), it is generally insufficient to separate the more closely related volcanic events on each island. Since there is considerable variation in the quantity, quality, and accessibility of specific obsidian sources, this knowledge is essential to a complete understanding of obsidian exploitation in this region (*17*).

For Sardinia, the exploitation of multiple sources was noted prior to the actual identification and characterization of each outcrop. The obsidian sources in the Monte Arci region of Sardinia have now been thoroughly documented and geochemically characterized (*18, 19*), while the results of recent fieldwork by this author on the other islands are expected to add significantly to earlier studies on Lipari (*20*), Palmarola (*21, 22*), and Pantelleria (*23, 24*). The western Mediterranean sources are briefly described here:

- Sardinian A. Type SA obsidian is black, glassy, and highly translucent. Abundant nodules up to 40 cm are typically found in soft perlitic matrices at Conca Cannas and Su Paris de Monte Bingias in the southwestern part of Monte Arci. Individual microlite crystals are visible in transmitted light, often with some flow orientation.
- Sardinian B1. Type SB1 is also black, but less glassy and usually opaque. It may be found as smaller nodules in harder matrices along the upper western flanks of Monte Arci at Punta Su Zippiri and Monte Sparau North, and as pyroclastic bombs up to 30cm in length on the slopes of Cuccuru Porcufurau.
- Sardinian B2. Type SB2 is black but ranges from virtually transparent to nearly opaque, and from very glassy to including many white phenocrysts up to 2mm in diameter. It also occurs along the western flanks of Monte Arci, but at lower elevations near Cucru Is Abis, Seddai, Conca S'Ollastu, and Bruncu Perda Crobina, occasionally in very large blocks.
- Sardinian C1. Type SC1 is black but frequently has well-defined external gray bands, and is less glassy and totally opaque. Rare pieces have red streaks or are partially transparent but tinted brown. This type may be found along the high ridge from Punta Pizzighinu to Perdas Urias on the northeastern side of Monte Arci, in blocks up to 20cm in length.
- Sardinian C2. Type SC2 is visually indistinguishable from SC1. It is commonly found as large blocks in secondary deposits between Perdas Urias and Santa Pinta. Since it is found mixed with material of SC1 type and differs chemically only in its trace concentrations of a few elements, the characterization of material as simply SC is probably sufficient for archaeological purposes.

- Lipari. Obsidian from Lipari is either glassy and very highly transparent, with a black to gray tint, or not very transparent due to the presence of many small phenocrysts which give it a gray streaky appearance. Historic-era volcanism has covered much of the island, but prehistorically available obsidian is still present in the Gabellotto gorge and along the northern and eastern coasts. There is no evidence that an earlier flow at Monte della Guardia was exploited in antiquity.
- Palmarola. Obsidian from Palmarola is almost always black and opaque, without any phenocrysts present. It is found primarily as secondary deposits along the southwestern coast of Monte Tramontana, typically in small nodules, and near Punta Vardella at the southeastern tip of the island, frequently in much larger blocks.
- Pantelleria - Balata dei Turchi. Nearly opaque dark-green obsidian is abundantly present in large blocks along the southern end of Pantelleria, with principal exposures at Balata dei Turchi and Salta La Vecchia. At least three chemically distinct volcanic layers have been identified.
- Pantelleria - Lago di Venere. Similar appearing obsidian of a slightly different composition occurs as smaller and infrequent nodules within perlitic matrices in the slopes above a natural hot spring in northeastern Pantelleria.
- Pantelleria - Gelkhamar. Obsidian is also thought to occur in western Pantelleria, based on the presence of a 5$^{th}$ chemical group among artifacts in this area. No geological material has been identified there, however, suggesting perhaps that it was available only as bombs in scattered secondary deposits which were mostly exhausted.

## Chemical Characterization of Obsidian Sources

Obsidians, by definition, have relatively limited ranges in their bulk element composition. For this reason, many Mediterranean and Near Eastern provenance studies have focused on trace element characterization of both geological samples and archaeological artifacts, primarily using x-ray fluorescence (*21, 23, 24, 25, 26, 27*) or instrumental neutron activation analysis (*14, 28, 29, 30, 31, 32, 33, 34*), although laser ablation ICP-MS is expected to become the technique of choice (*35, 36, 37*). Nevertheless, quantitative analysis of the major and minor elements is sufficient to differentiate not only the different island sources (Sardinia, Lipari, Palmarola, and Pantelleria, as well as Melos and Giali in the Aegean) but even several sub-sources on Sardinia, Pantelleria and Melos, so that at least in the Mediterranean all currently known and useful source distinctions can be made without resorting to trace element techniques.

The most complete characterizations currently available for the western Mediterranean obsidian sources have been accomplished using XRF and the electron microprobe (Table 1). Comparison of the average major/minor element values for the western Mediterranean sources reveals that $K_2O$ concentration alone can often distinguish SA from SB from SC (5.24 ± 0.12; 5.48 ± 0.17; and 5.89 ± 0.34 %, respectively); SB1 has 1.2% lower $SiO_2$ and 0.7% higher $Al_2O_3$ than SB2; and type SC obsidian is distinguished from all other western Mediterranean obsidians by its high $Al_2O_3$ (13.9%), MnO (0.14%), and BaO (0.11%). All four of these types are easily distinguished from Lipari and Palmarola obsidian by their $Na_2O$ concentrations (<3.5 vs. >4%), while the peralkaline obsidian from Pantelleria is easily distinguished by its extremely low $SiO_2$, $Al_2O_3$, and CaO, and its extremely high $Fe_2O_3$ and $Na_2O$ concentrations (Figure 2).

## Early Neolithic Obsidian Artifacts

Sixty-one obsidian artifacts from four Early Neolithic sites were selected for chemical analysis. The analysis of significant-sized assemblages from single cultural periods allows direct comparison between contemporary sites in different geographic locations. Many early provenance studies, however, relied on very few artifacts from individual sites so that such patterns in the exploitation of specific obsidian sources could not be determined (*15*).

**Table I. Electron microprobe data for Sardinian obsidian**

| *Type* | *$SiO_2$* | *$Al_2O_3$* | *$TiO_2$* | *$Fe_2O_3$* | *MgO* | *CaO* | *$Na_2O$* | *$K_2O$* | *MnO* | *$P_2O_5$* | *BaO* | *Total* |
|---|---|---|---|---|---|---|---|---|---|---|---|---|
| *Sardinia A* | | | | | | | | | | | | |
| ave | 74.72 | 13.40 | 0.09 | 1.25 | 0.08 | 0.59 | 3.44 | 5.26 | 0.08 | 0.06 | 0.02 | 99.00 |
| sd | 0.26 | 0.15 | 0.01 | 0.09 | 0.01 | 0.04 | 0.16 | 0.22 | 0.01 | 0.01 | 0.02 | |
| *Sardinia B1* | | | | | | | | | | | | |
| ave | 73.87 | 13.63 | 0.17 | 1.33 | 0.12 | 0.75 | 3.38 | 5.55 | 0.10 | 0.04 | 0.05 | 99.00 |
| sd | 0.35 | 0.10 | 0.03 | 0.20 | 0.04 | 0.06 | 0.10 | 0.20 | 0.02 | 0.01 | 0.02 | |
| *Sardinia B2* | | | | | | | | | | | | |
| ave | 75.05 | 12.97 | 0.13 | 1.17 | 0.11 | 0.57 | 3.34 | 5.51 | 0.08 | 0.04 | 0.02 | 99.00 |
| sd | 0.33 | 0.15 | 0.02 | 0.17 | 0.02 | 0.02 | 0.22 | 0.35 | 0.01 | 0.01 | 0.02 | |
| *Sardinia C* | | | | | | | | | | | | |
| ave | 72.71 | 13.92 | 0.27 | 1.53 | 0.21 | 0.88 | 3.30 | 5.90 | 0.14 | 0.03 | 0.11 | 99.00 |
| sd | 0.37 | 0.19 | 0.03 | 0.21 | 0.07 | 0.10 | 0.20 | 0.30 | 0.01 | 0.01 | 0.03 | |

NOTE: Mean and standard deviation in percent. Values standardized to a total of 99% allowing 1% for water and trace elements.

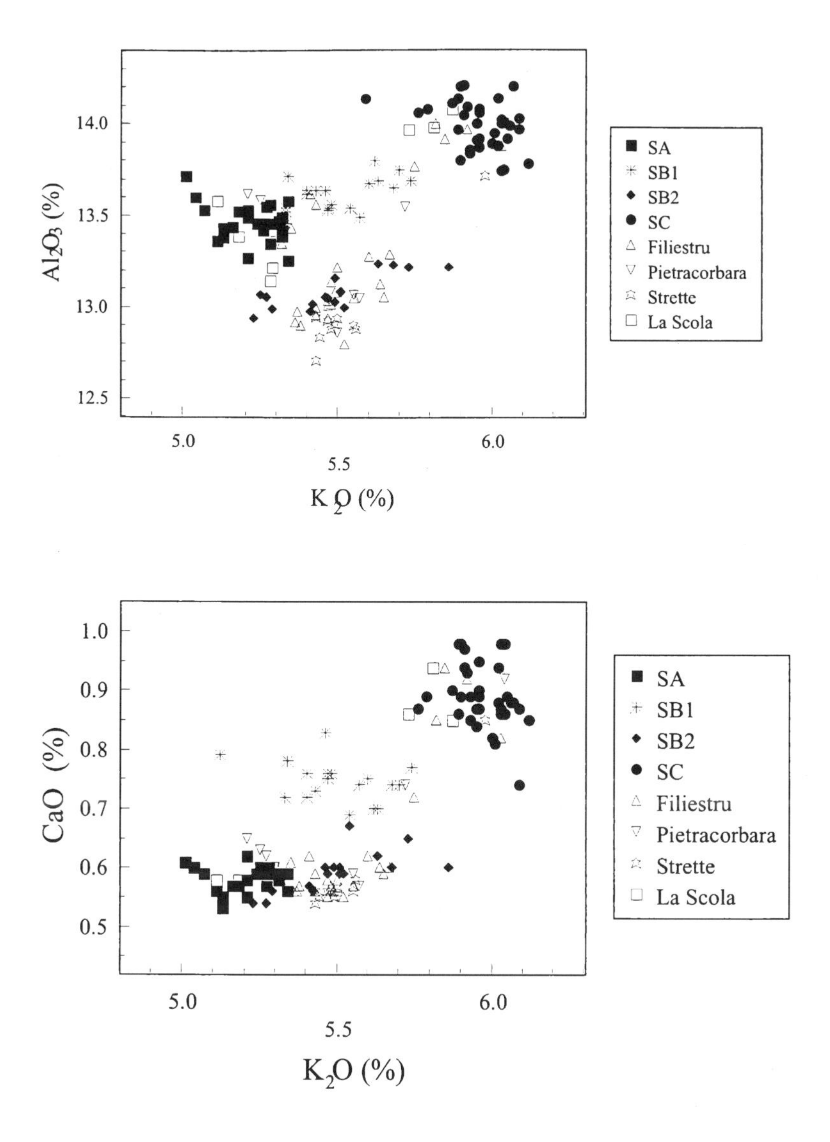


*Figure 2. Bivariate plots of major element oxides for Sardinian obsidian sources and artifacts analyzed in this study.*
*$CaO$ vs. $K_2O$* (top); *$Al_2O_3$ vs. $K_2O$* (bottom)

Obsidian accounts for 17% of the over 400 Cardial Early Neolithic stone tools excavated at Filiestru Cave in northwestern Sardinia (*3, 38*). Twenty-seven obsidian artifacts were selected from these levels in trenches B (B10; B11 t.2 and t.3) and D (stratigraphic column) for chemical analysis. An additional 45 obsidian artifacts from the same contexts were also visually analyzed, while obsidian artifacts from a later Early Neolithic (Filiestru culture) level as well as Middle and Late Neolithic contexts were also analyzed by chemical and visual means (*17*).

Ten obsidian artifacts each were also selected from two sites in northern Corsica. The Cardial Early Neolithic levels at Strette included over 1000 lithic artifacts, with over 11% in obsidian; the ten artifacts selected came from level XIV of Strette 1 (*39, 40*). The 24 obsidian artifacts excavated in level VI at Pietracorbara, however, comprised only about 6% of the lithic assemblage there (*41*).

Fourteen obsidian artifacts from the small islet of La Scola near Pianosa in the Tuscan Archipelago were also analyzed. About 20% of the 1500 lithics recovered are in obsidian, and they are predominantly microliths with no retouch and showing only the faintest traces of use (*42, 43*). The analyzed obsidian was recovered from units BC 11-12 I/3; A12 tg3 (I3); and A8-9-10 B12, all associated with Cardial ceramics.

## Chemical Analyis of Obsidian Artifacts

Electron probe microanalysis using wavelength dispersive x-ray spectrometers was selected as the method of choice for analyzing large numbers of archaeological artifacts since only a tiny 1-2mm sample needs to be removed, sample preparation is minimal, and fully quantitative measurement of all major and minor elements is possible at a very low per-sample cost. Although recent studies indicate that energy dispersive spectrometry can also produce excellent quantitative results for ancient glassy materials (*44, 45*), $TiO_2$, MgO, MnO, $P_2O_5$, and BaO are below the minimum detection limits of SEM-EDS systems, and at least $TiO_2$, MgO, and BaO are important discriminators among western Mediterranean obsidian sources (*19*).

Cylinders one inch in diameter were made using Epotek two-part epoxy, and up to 18 holes 2mm in diameter and 3mm in depth were drilled in the flat surface of the hardened disk. Fresh epoxy and obsidian samples, cut earlier with a fast-speed diamond saw if necessary, were inserted in the holes and allowed to harden for 48 hours before grinding and polishing to a 1-micron finish. Sample disks were carbon-coated prior to analysis to minimize local surface charging under the electron beam.

Samples were analyzed using a Cameca MBX electron microprobe equipped with a Tracor Northern computer control system, and quantitative wavelength dispersive analysis was performed using a modified version of Sandia TASK8.

Samples were excited by a wide 40 micron electron beam at 15 KeV in order to minimize the influence of tiny microlite inclusions (typically high in iron) on the overall composition of the sample, as well as to prevent the heat-induced decomposition of alkali elements during analysis. Two to three points on each sample were analyzed to further reduce any heterogeneity in the results. The X-radiation produced was measured by wavelength dispersive spectrometry (WDS) using TAP (thallium acid phthalate), PET (penta-erythritol) and LiF (lithium fluoride) crystals, with counting times of 10-80 seconds per element. All samples were analyzed for $SiO_2$, $Al_2O_3$, $TiO_2$, $Fe_2O_3$, MgO, CaO, $Na_2O$, $K_2O$ and BaO; MnO, $P_2O_5$ were also measured for some samples. The measured X-ray intensities were corrected for matrix effects, absorption, and secondary fluorescence by the Bence-Albee correction program. Results were internally calibrated against international standards, and a laboratory reference material (hornblende) was repeatedly measured to insure consistency between analytical sessions. The results for each point analysis were normalized to total 99.00% (allowing 1% for water and trace elements) and then the average calculated for each sample (Table 2).

Seven of the artifacts from the La Scola site were analyzed by ICP-MS. Samples were ultrasonically cleaned and dried, and then pulverized at liquid nitrogen temperatures using a SPEX 6700 freezer mill. 100 mg of the powdered sample were mixed with 0.2 mL aqua regia and 2.5 mL hydrofluoric acid and dissolved in a Parr acid digestion bomb at 120 °C for 2 hours. Samples were then diluted to a total volume of 250 ml so that elements of interest would be present at concentrations of 500 ppb or less.

Immediately prior to analysis, 100 μL of a 10 ppm $^{115}In$ solution were added to a 10 mL aliquot of each sample as an internal spike to correct for instrument drift. Analyses were performed on a Fisons PQ 2 Plus quadrupole ICP mass spectrometer. Rather than produce calibration curves for each of the 41 elements analyzed, simple blank subtraction was used and the results for the artifacts were compared directly against geological source samples prepared and analyzed in the same way (*35*).

## Results and Discussion

Source attributions were initially made using bivariate element plots, and were confirmed by applying multivariate discriminant functions determined for the geological samples to the artifact samples using the jackknife option of the software program BMDP 7M. The specific geological source of all 61 artifacts analyzed was identified with greater than 95% statistical probability (Table 3). Not surprisingly, all of the artifacts from all four sites come from the Monte Arci sources in Sardinia. To date, no obsidian from any of the other island sources has been found on Corsica or Sardinia. Obsidian from both Lipari and Palmarola,

## Table II. Electron Microprobe Data for Obsidian Artifacts

| *No.* | *$SiO_2$* | *$Al_2O_3$* | *$TiO_2$* | *$Fe_2O_3$* | *MgO* | *CaO* | *$Na_2O$* | *$K_2O$* | *MnO* | *$P_2O_5$* | *BaO* | *Total* |
|---|---|---|---|---|---|---|---|---|---|---|---|---|
| *La Scola* | | | | | | | | | | | | |
| 292 | 74.85 | 13.58 | 0.07 | 1.19 | 0.09 | 0.58 | 3.35 | 5.11 | | | 0.03 | 98.85 |
| 293 | 72.82 | 13.97 | 0.26 | 1.55 | 0.26 | 0.86 | 3.19 | 5.73 | | | 0.15 | 98.85 |
| 294 | 73.27 | 14.08 | 0.26 | 1.10 | 0.06 | 0.85 | 3.19 | 5.87 | | | 0.14 | 98.85 |
| 295 | 74.88 | 13.39 | 0.06 | 1.17 | 0.08 | 0.58 | 3.37 | 5.18 | | | 0.01 | 98.85 |
| 296 | 75.22 | 13.15 | 0.12 | 1.17 | 0.12 | 0.59 | 3.26 | 5.28 | 0.11 | 0.07 | 0.02 | 99.00 |
| 297 | 75.21 | 13.22 | 0.11 | 0.92 | 0.09 | 0.60 | 3.36 | 5.29 | | | 0.05 | 98.85 |
| 298 | 72.96 | 13.98 | 0.32 | 1.25 | 0.11 | 0.94 | 3.20 | 5.81 | | | 0.17 | 98.85 |
| *Grotta Filiestru* | | | | | | | | | | | | |
| 1170 | 75.27 | 12.94 | 0.11 | 1.08 | 0.10 | 0.55 | 3.35 | 5.47 | 0.07 | 0.04 | 0.01 | 99.00 |
| 1171 | 74.30 | 13.56 | 0.10 | 1.33 | 0.08 | 0.59 | 3.45 | 5.43 | 0.08 | 0.07 | 0.02 | 99.00 |
| 1172 | 74.62 | 13.22 | 0.13 | 1.25 | 0.15 | 0.58 | 3.42 | 5.50 | 0.06 | 0.04 | 0.01 | 99.00 |
| 1173 | 74.42 | 13.29 | 0.14 | 1.23 | 0.14 | 0.60 | 3.42 | 5.67 | 0.07 | 0.02 | 0.01 | 99.00 |
| 1174 | 74.41 | 13.62 | 0.09 | 1.19 | 0.07 | 0.62 | 3.37 | 5.41 | 0.08 | 0.07 | 0.04 | 99.00 |
| 1175 | 75.06 | 13.04 | 0.13 | 1.12 | 0.10 | 0.55 | 3.39 | 5.47 | 0.07 | 0.03 | 0.02 | 99.00 |
| 1176 | 73.85 | 13.77 | 0.17 | 1.04 | 0.07 | 0.72 | 3.42 | 5.75 | 0.11 | 0.03 | 0.07 | 99.00 |
| 1177 | 75.06 | 13.00 | 0.12 | 1.11 | 0.11 | 0.56 | 3.42 | 5.43 | 0.08 | 0.04 | 0.00 | 99.00 |
| 1178 | 75.11 | 13.02 | 0.12 | 1.07 | 0.10 | 0.57 | 3.43 | 5.48 | 0.06 | 0.04 | 0.01 | 99.00 |
| 1179 | 75.01 | 13.05 | 0.12 | 1.13 | 0.09 | 0.57 | 3.36 | 5.55 | 0.08 | 0.04 | 0.00 | 99.00 |
| 1180 | 74.97 | 12.96 | 0.12 | 1.14 | 0.13 | 0.56 | 3.39 | 5.43 | 0.08 | 0.03 | 0.04 | 99.00 |
| 1181 | 74.96 | 13.01 | 0.12 | 1.17 | 0.11 | 0.58 | 3.45 | 5.47 | 0.08 | 0.03 | 0.02 | 99.00 |
| 1182 | 72.71 | 13.88 | 0.26 | 1.55 | 0.21 | 0.82 | 3.29 | 6.03 | 0.12 | 0.03 | 0.10 | 99.00 |
| 1183 | 74.90 | 13.08 | 0.13 | 1.15 | 0.11 | 0.56 | 3.43 | 5.49 | 0.08 | 0.04 | 0.03 | 99.00 |
| 1184 | 72.37 | 13.92 | 0.29 | 1.58 | 0.22 | 0.94 | 3.31 | 5.85 | 0.17 | 0.03 | 0.15 | 99.00 |
| 1185 | 74.35 | 13.28 | 0.15 | 1.32 | 0.14 | 0.62 | 3.41 | 5.60 | 0.08 | 0.05 | 0.00 | 99.00 |
| 1186 | 74.97 | 13.14 | 0.12 | 0.99 | 0.12 | 0.57 | 3.48 | 5.48 | 0.08 | 0.04 | 0.03 | 99.00 |
| 1227 | 75.31 | 12.92 | 0.13 | 1.18 | 0.10 | 0.56 | 3.30 | 5.36 | 0.07 | 0.05 | 0.01 | 99.00 |
| 1228 | 74.81 | 13.35 | 0.10 | 1.24 | 0.07 | 0.58 | 3.39 | 5.32 | 0.08 | 0.06 | 0.01 | 99.00 |
| 1229 | 74.71 | 13.13 | 0.16 | 1.14 | 0.14 | 0.60 | 3.34 | 5.64 | 0.08 | 0.03 | 0.03 | 99.00 |
| 1230 | 75.46 | 12.80 | 0.11 | 1.09 | 0.07 | 0.55 | 3.28 | 5.52 | 0.07 | 0.03 | 0.02 | 99.00 |
| 1231 | 72.67 | 13.97 | 0.28 | 1.60 | 0.15 | 0.92 | 3.31 | 5.92 | 0.15 | 0.02 | 0.11 | 99.00 |
| 1232 | 75.28 | 12.98 | 0.12 | 1.11 | 0.10 | 0.56 | 3.35 | 5.37 | 0.08 | 0.03 | 0.00 | 99.00 |
| 1233 | 74.61 | 13.43 | 0.09 | 1.31 | 0.09 | 0.61 | 3.40 | 5.35 | 0.06 | 0.04 | 0.00 | 99.00 |
| 1234 | 75.04 | 13.06 | 0.15 | 1.04 | 0.12 | 0.59 | 3.26 | 5.65 | 0.07 | 0.04 | 0.01 | 99.00 |
| 1235 | 72.49 | 14.00 | 0.27 | 1.43 | 0.18 | 0.85 | 3.26 | 5.82 | 0.14 | 0.03 | 0.08 | 99.00 |
| 1236 | 75.35 | 12.90 | 0.11 | 1.16 | 0.09 | 0.57 | 3.32 | 5.38 | 0.07 | 0.03 | 0.02 | 99.00 |

**Table II (continued). Electron Microprobe Data for Obsidian Artifacts**

| *No.* | *$SiO_2$* | *$Al_2O_3$* | *$TiO_2$* | *$Fe_2O_3$* | *MgO* | *CaO* | *$Na_2O$* | *$K_2O$* | *MnO* | *$P_2O_5$* | *BaO* | *Total* |
|---|---|---|---|---|---|---|---|---|---|---|---|---|
| *Pietracorbara* | | | | | | | | | | | | |
| 1308 | 74.61 | 13.36 | 0.09 | 1.28 | 0.09 | 0.60 | 3.53 | 5.30 | | | 0.01 | 98.85 |
| 1309 | 74.63 | 13.42 | 0.09 | 1.30 | 0.08 | 0.59 | 3.43 | 5.31 | | | 0.00 | 98.85 |
| 1310 | 75.23 | 12.86 | 0.11 | 1.11 | 0.10 | 0.55 | 3.38 | 5.50 | | | 0.00 | 98.85 |
| 1311 | 74.36 | 13.59 | 0.10 | 1.19 | 0.07 | 0.63 | 3.62 | 5.25 | | | 0.02 | 98.85 |
| 1312 | 74.14 | 13.62 | 0.09 | 1.35 | 0.07 | 0.65 | 3.71 | 5.21 | | | 0.03 | 98.85 |
| 1313 | 74.81 | 13.07 | 0.14 | 1.15 | 0.12 | 0.59 | 3.39 | 5.55 | | | 0.04 | 98.85 |
| 1314 | 72.61 | 14.00 | 0.30 | 1.31 | 0.19 | 0.92 | 3.39 | 6.04 | | | 0.08 | 98.85 |
| 1315 | 74.59 | 13.42 | 0.08 | 1.19 | 0.07 | 0.62 | 3.57 | 5.27 | | | 0.06 | 98.85 |
| 1316 | 73.44 | 13.55 | 0.18 | 1.60 | 0.18 | 0.74 | 3.43 | 5.72 | | | 0.03 | 98.85 |
| 1317 | 74.64 | 13.05 | 0.14 | 1.22 | 0.14 | 0.57 | 3.46 | 5.57 | | | 0.06 | 98.85 |
| *Strette* | | | | | | | | | | | | |
| 1330 | 75.09 | 12.91 | 0.12 | 1.14 | 0.11 | 0.55 | 3.41 | 5.49 | | | 0.04 | 98.85 |
| 1331 | 75.01 | 12.88 | 0.14 | 1.12 | 0.12 | 0.58 | 3.39 | 5.56 | | | 0.07 | 98.85 |
| 1332 | 75.17 | 12.95 | 0.12 | 1.10 | 0.11 | 0.54 | 3.42 | 5.43 | | | 0.03 | 98.85 |
| 1333 | 75.15 | 12.88 | 0.13 | 1.13 | 0.09 | 0.56 | 3.41 | 5.48 | | | 0.03 | 98.85 |
| 1334 | 75.14 | 12.94 | 0.14 | 1.08 | 0.11 | 0.57 | 3.36 | 5.50 | | | 0.01 | 98.85 |
| 1335 | 74.79 | 12.71 | 0.13 | 1.74 | 0.09 | 0.56 | 3.39 | 5.43 | | | 0.02 | 98.85 |
| 1336 | 75.17 | 12.93 | 0.12 | 1.15 | 0.09 | 0.56 | 3.36 | 5.47 | | | 0.00 | 98.85 |
| 1337 | 75.09 | 12.90 | 0.12 | 1.09 | 0.10 | 0.56 | 3.40 | 5.55 | | | 0.05 | 98.85 |
| 1338 | 72.89 | 13.72 | 0.25 | 1.54 | 0.20 | 0.85 | 3.35 | 5.98 | | | 0.08 | 98.85 |
| 1339 | 75.32 | 12.84 | 0.11 | 1.09 | 0.11 | 0.55 | 3.40 | 5.44 | | | 0.00 | 98.85 |

NOTE: All units in percent. Values standardized to a total of 99% allowing 1% for water and trace elements, or to 98.85% when MnO and $P_2O_5$ were not determined.

**Table III. Source Attributions of Archaeological Artifacts**

| *Site* | *SA* | *SB1* | *SB2* | *SC* | *Total* |
|---|---|---|---|---|---|
| La Scola | 3 | 0 | 5 | 6 | 14 |
| Filiestru | 4 | 1 | 18 | 4 | 27 |
| Pietracorbara | 5 | 1 | 3 | 1 | 10 |
| Strette | 0 | 0 | 9 | 1 | 10 |
| *total* | 12 | 2 | 35 | 12 | 61 |

however, have been identified at the Early Neolithic site of Cala Giovanna on Pianosa, in sight of and presumably contemporary with La Scola, although Sardinian obsidian does account for more than 90% of the obsidian found there (*27, 46*).

The relative proportions of each Sardinian source are also significant. Types SA and SC each represent about 20% of the artifacts tested, but are not consistently present in quantity at all four sites, type SA being absent at Strette and type SC accounting for only two artifacts total at the two Corsican sites. Type SB1 is represented by just two artifacts, but notably at two different sites, while type SB2 is consistently well represented at all four sites, accounting for nearly 60% of all the artifacts tested.

This pattern is consistent with that observed at other Early Neolithic sites in Corsica (*27*), Tuscany (*47*), and Liguria (*34*), but contrasts sharply with southern France (*26, 29*) where type SA accounts for the vast majority of obsidian found there (Figure 3). These patterns would not have been observable without the level of resolution enabled by the analytical methodology used here, and allow us to make interpretations that were not possible using more basic distributional data (*48, 49*). Down-the-line trade (*50*), characteristic of societies without marked social hierarchies, is widely considered to be a common exchange mechanism during the Neolithic period. Settlements close to an obsidian source would have direct access and an abundant supply. They would have exchanged some of their obsidian with neighboring villages further from this supply zone, and these villages would have done the same with others even further away, so that the frequency of obsidian would decrease with distance from the source. The presence, in all but one case, of three Monte Arci obsidian types, with a high importance of type SB2 in all cases, at Filiestru, Strette, Pietracorbara, Lumaca, La Scola, Cala Giovanna, Paduletto di Castagneto, Querciolaia, and Arene Candide, is consistent with such a model for the

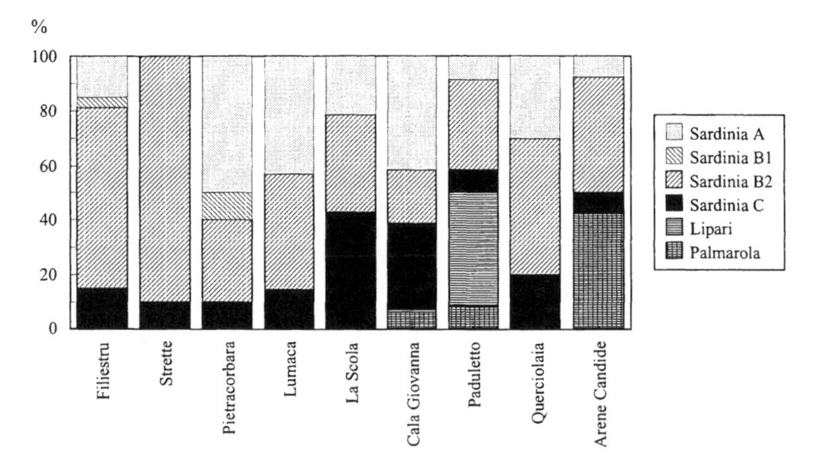


*Figure 3. Relative frequency of obsidian at Early Neolithic sites in Sardinia, Corsica, and northern Italy. Other authors have not separated any of the SB subsources, but are most likely to be SB2 as they are shown here.*

distribution of obsidian from Sardinia to Corsica to Tuscany and northwards to Liguria, perhaps in exchange for ground stone tools, pottery, and more perishable materials (*1*).

The obviously different pattern in southern France implies, however, that more than one exchange mechanism may have been in place at one time, and/or that cultural and other factors were much more complex than the heuristic explanation proposed here. Maritime contacts between Sardinia and the mainland perhaps were not necessarily routed across the shortest open-water crossings. Furthermore, there may have been significant differences in the use-function of obsidian within different groups, perhaps depending on the availability of alternative lithic resources. The integration of provenance studies with typological and use-wear analysis (*51*) shows great promise for testing these hypotheses.

In conclusion, high-resolution provenance studies of western Mediterranean obsidian artifacts produce more detailed information on Early Neolithic exchange patterns than can studies which only attribute artifacts at the island level. This was a time when an agricultural way of life was first introduced and the Mediterranean islands were first widely settled. Obsidian, as the most visible indicator in the archaeological record of intercultural contact, is a critical material for the reconstruction of this most important transition in human prehistory.

## References


1. Tykot, R.H. In *Social Dynamics of the Prehistoric Central Mediterranean;* Tykot, R H.; Morter, J.; Robb, J. E., Eds.; Accordia Research Center: London, 1999; pp 67-82.
2. Tykot, R. H. In *Radiocarbon Dating and Italian Prehistory;* Skeates, R.; Whitehouse, R., Eds.; Accordia Research Centre: London, 1994; pp 115-145.
3. Trump, D. H. *La grotta di Filiestru a Bonu Ighinu, Mara.* Quaderni 13; Dessì: Sassari, Italy, 1983.
4. Barnett, W. K. *Antiquity* **1990**, *64*, 859-865.
5. Chapman, J.C. *Journal of Mediterranean Archaeology* **1988**, *1*, 3-25.
6. Barnett, W. K. In *Europe's First Farmers*; Price, T.D., Ed.; Cambridge University Press: Cambridge, England, 2000; pp 93-116.
7. Binder, D. In *Europe's First Farmers*; Price, T. D., Ed.; Cambridge University Press: Cambridge, England, 2000; pp 117-143.
8. Brandaglia, M. *Studi per l'ecologia del Quaternario* **1985**, *7*, 53-76.
9. Binder, D. *Le Néolithique Ancien Provençal. Typologie et Technologie des Outillages Lithiques*; Gallia Préhistoire Supplement 24; Paris: C.N.R.S., 1987.
10. Costa, L.; Sicurani, J. In *Il primo popolamento Olocenico dell'area corso-toscana*; Tozzi, C.; Weiss, M. C., Eds.; Edizioni ETS: Pisa, Italy, 2000; pp 189-200.
11. de Lanfranchi, F. *Bulletin de la Société préhistorique française* **1980**, *77(4)*, 123-28.
12. de Lanfranchi, F. *Corsica Antica* **1993**, 2-9.
13. Cann, J. R.; Renfrew, C. *Proceedings of the Prehistoric Society* **1964**, *30*, 111-33.
14. Hallam, B. R.; Warren, S. E.; Renfrew, C. *Proceedings of the Prehistoric Society* **1976**, *42*, 85-110.
15. Tykot, R. H.; Ammerman, A. J. *Antiquity* **1997**, *274*, 1000-1006.
16. Bigazzi, G.; Radi, G. In *XIII International Congress of Prehistoric and Protohistoric Sciences-Forlí-Italia-8/14 September 1996*; Arias, C.; Bietti, A.; Castelletti, L.; Peretto, C., Eds.; A.B.A.C.O.: Forlí, Italy, 1996; Vol. 1, pp 149-156.
17. Tykot, R. H. *J. Mediterranean Archaeology* **1996**, *9*, 39-82.
18. Tykot, R. H. In *Sardinia in the Mediterranean: A Footprint in the Sea*; Tykot, R. H.; Andrews, T. K., Eds.; Sheffield Academic Press: Sheffield, England, 1992; pp 57-70.
19. Tykot, R. H. *J. Archaeol. Sci.* **1997**, *24,* 467-479.
20. Pichler, H. *Rendiconti Societa Italiana di Mineralogia e Petrologia* **1980**, *36*, 415-440.
21. Herold, G. Ph.D. thesis, Universität (TH) Fridericiana Karlsruhe, Germany, 1986.

22. Barberi, F.; Borsi, S.; Ferrara, G.; Innocenti, F. *Memorie della Societa Geologica Italiana* **1967**, *6*, 581-606.
23. Francaviglia, V. *Preistoria Alpina - Museo Tridentino di Scienze Naturali* **1986**, *20*, 311-32.
24. Francaviglia, V. *J. Archaeol. Sci.* **1988**, *15*, 109-22.
25. Francaviglia, V.; Piperno, M. *Revue d'Archéométrie* **1987**, *11*, 31-39.
26. Crisci, G. M.; Ricq-de Bouard, M.; Lanzaframe, U.; de Francesco, A.M. *Gallia Préhistoire* **1994**, *36*, 299-327.
27. De Francesco, A. M.; Crisci, G. M. In *Il primo popolamento Olocenico dell'area corso-toscana;* Tozzi, C.; Weiss, M. C., Eds.; Edizioni ETS: Pisa, Italy, 2000; pp 253-258.
28. Williams-Thorpe, O.; Warren, S. E.; Barfield, L. H. *Preistoria Alpina - Museo Tridentino di Scienze Naturali* **1979**, *15*, 73-92.
29. Williams-Thorpe, O.; Warren, S. E.; Courtin, J. *J. Archaeol. Sci.* **1984**, *11*, 135-46.
30. Crummett, J.G.; Warren, S. E. In *The Acconia Survey: Neolithic Settlement and the Obsidian Trade*; Ammerman, A. J., Ed.; Institute of Archaeology Occ. Pub. 10; Institute of Archaeology: London, 1985; pp 107-114.
31. Ammerman, A. J.; Cesana, A.; Polglase, C.; Terrani, M. *J. Archaeol. Sci.* **1990**, *17*, 209-20.
32. Renfrew, C.; Aspinall, A. In *Les Industries Lithiques Taillées de Franchthi (Argolide, Grèce). Tome II. Les Industries du Mésolithique et du Néolithique Initial*; Perlès, C., Ed.; Indiana University Press: Bloomington, IN, 1990; pp 257-70.
33. Randle, K.; Barfield, L. H.; Bagolini, B. *J. Archaeol. Sci.* **1993**, *20*, 503-509.
34. Ammerman, A. J.; Polglase, C. R. In *Arene Candide: A Functional and Environmental Assessment of the Holocene Sequence (Excavations Bernabo Brea-Cardini 1940-50)*; Maggi, R., Ed.; Il Calamo: Rome, Italy, 1997; pp 573-592.
35. Tykot, R. H.; Young, S. M. M. In *Archaeological Chemistry V*; Orna, M.V., Ed.; American Chemical Society: Washington, DC, 1996; pp 116-130.
36. Gratuze, B. *J. Archaeol. Sci.* **1999**, *26*, 869-881.
37. Halliday, A. N.; Der-Chuen, L.; Christensen, J. N.; Rehkamper, M.; Yi, W.; Luo, X.; Hall, C. M.; Ballentione, C. J.; Pettke, T.; Stirling, C. *Geochim. Cosmochim. Acta* **1998**, *62*, 919-940.
38. Trump, D. H. In *Studies in Sardinian Archaeology*; Balmuth, M.S.; Rowland, R. J. Jr., Eds.; University of Michigan Press: Ann Arbor, MI, 1984; pp 1-22.
39. Magdeleine, J. *Archeologica Corsa* **1983-84**, *8-9*, 30-50.
40. Magdeleine, J.; Ottaviani, J. C. *Bulletin de la Societe des Sciences Historiques & Naturelles de la Corse* **1986**, *650*, 61-90.
41. Magdeleine, J. *Bulletin de la Société Préhistorique Française* **1995**, *92(3)*, 363-377.

42. Ducci, S.; Perazzi, P. In *XIII International Congress of Prehsitoric and Protohistoric Sciences-Forlí-Italia-8/14 September 1996*; Cremonesi, R. G.; Tozzi, C.; Vigliardi, A.; Peretto, C., Eds.; A.B.A.C.O.: Forlí, Italy, 1998; Vol. 3, 425-430.
43. Ducci, S.; Guerrini, M. V., Perazzi, P. In *Il primo popolamento Olocenico dell'area corso-toscana*; Tozzi, C.; Weiss, M. C., Eds.; Edizioni ETS: Pisa, Italy, 2000; pp 83-90.
44. Verità, M.; Basso, R.; Wypyski, M. T.; Koestler, R. J. *Archaeometry* **1994**, *36*, 241-251.
45. Acquafredda, P.; Andriani, T.; Lorenzoni, S.; Zanettin, E. *J. Archaeol. Sci.* **1999**, *26*, 315-325.
46. Bonato, M.; Lorenzi, F.; Nonza, A.; Radi, G.; Tozzi, C.; Weiss, M. C.; Zamagni, B. In *Il primo popolamento Olocenico dell'area corso-toscana*; Tozzi, C.; Weiss, M. C., Eds.; Edizioni ETS: Pisa, Italy, 2000; pp 91-116.
47. Tykot, R. H.; IIPP
48. Pollmann, H.-O. *Obsidian im Nordwestmediterranean Raum: Seine Verbreitung und Nutzung im Neolithikum und Äneolithikum*; BAR International Series 585. Tempus Reparatvm: Oxford, England, 1993.
49. Renfrew, C. *Current Anthropology* **1969**, *10*, 151-160.
50. Tykot, R. H. In *Written in Stone: The Multiple Dimensions of Lithic Analysis*; Kardulias, P.N.; Yerkes, R., Eds.; Lexington Books: Baltimore, MD, 2002.
51. Hurcombe, L.; Phillips, P. In *Sardinian and Aegean Chronology: Towards the Resolution of Relative and Absolute Dating in the Mediterranean;* Balmuth, M. S.; Tykot, R. H., Eds; Oxbow Books: Oxford, England, 1998; pp 93-102.

## Chapter 12

# Chemical, Technological, and Social Aspects of Pottery Manufacture in the La Quemada Region of Northwest Mexico

**E. Christian Wells and Ben A. Nelson**

**Department of Anthropology, Arizona State University, Tempe, AZ 85287**

Using scanning electron microprobe analysis and simple re-firing experiments, this study reconstructs the operational sequence of production and decoration of incised-engraved pottery sherds from the prehispanic site of La Quemada, Zacatecas, Mexico. The results of these analyses suggest that certain steps in the firing and decorative processes were manipulated by local producers over time, which resulted in some of the conditions conducive to craft specialization.

Reconstructing the process of ceramic production has long been a focus of archaeological research concerned with the underlying practices involved in technological innovation, as well as the forces shaping cultural traditions. Over the past century, archaeological chemistry unquestionably has made significant contributions to these research objectives. Chemical studies alone, however, can seldom be used to answer anthropological questions about the relationship between pottery manufacture and technological innovation. Instead, archaeologists must employ multiple lines of evidence from independent datasets to evaluate critically information produced from chemical characterizations. In this paper, we present the results of a series of chemical and physical analyses to reconstruct the operational sequence of production of incised-engraved pottery vessels from the Malpaso Valley in south-central Zacatecas, Mexico, manufactured roughly AD 500 to 900, with the greater goal of elucidating how ceramic technology articulated with the local craft economy. To this end, three basic questions about the production process are addressed: 1) What raw material sources were utilized?, 2) How were the vessels formed and decorated?, and 3) What techniques were involved in the firing process? The results of this study suggest that certain steps in the firing and decorative processes were manipulated increasingly by local producers. These efforts resulted in significant technological innovations, specifically, the establishment of a consistent firing temperature and a shift to a post-fire engraving procedure that required the decoration of only those vessels that survived the sintering process. More broadly, these findings inform about some of the ways in which technological innovations can result in conditions that are conducive to craft specialization in middle-range societies.

## Geological and Archaeological Context of the Malpaso Valley

The Malpaso Valley is a relatively narrow floodplain (ca. 2000 m asl), approximately 10 km wide, and is currently cut on its easternmost edge by the Malpaso River (Figure 1). The region is bounded on the east and west by foothills, which give way to the Sierra Fría to the east and the Sierra Madre Occidental to the west. The valley floor adjacent to the main course of the Malpaso River is composed of alluvium that was deposited by lateral migrations of the river channel and by the deposition of fine-grained sediments deposited when river waters periodically covered the floodplain. A sedimentary conglomerate characterizes the eastern side of the valley extending from the valley's northern extreme to the southwestern portion of Villanueva. The Miocene extrusion of volcanic material in the valley resulted in the deposition of rhyolitic flows and tuffs, andesites, and basalts, the latter of which are restricted to the eastern and western valley margins. The rock-forming minerals found in these predominant extrusive igneous rock types include quartz, plagioclase feldspar, biotite, and olivine. These minerals occur to varying degrees in Malpaso Valley pottery (1), as well as in naturally occurring inclusions in local outcroppings of clay-rich sediments, which can be found eroding out of arroyo cuts into the alluvial floodplain and the surrounding foothills.

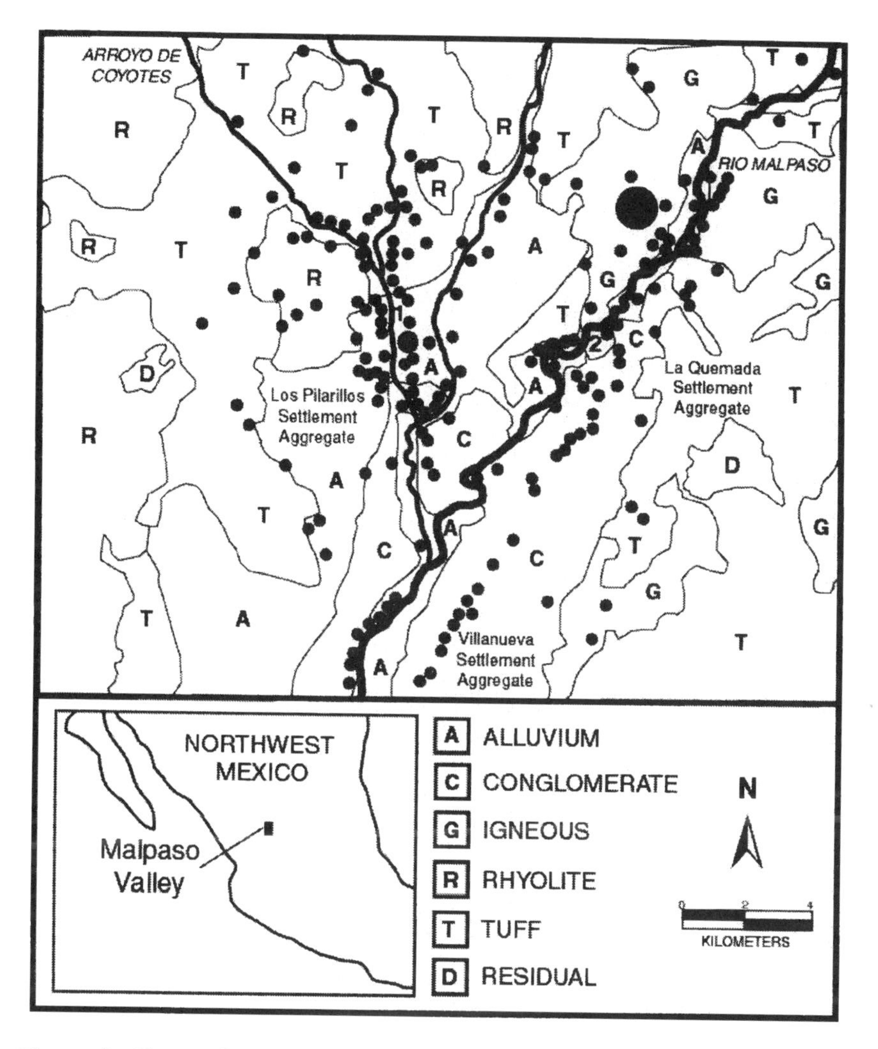


Figure 1. The geology of the Malpaso Valley, showing the locations of La Quemada (the largest dot in the northeast quadrant), outlying settlements (represented by smaller dots), and the clay outcroppings (numbers 1 and 2).

The valley contains three main archaeological zones, the largest of which includes the site of La Quemada, a sizeable civic-ceremonial center situated roughly 250 m above the valley floor on a prominent hilltop at the north end of the valley. The site is composed of a monumental core that contains a central ballcourt, a large colonnaded hall, at least 12 pyramids, and a series of sunken-patio compounds that were probably elite residences (2). In addition, there are more than 56 artificial terraces that flank the upper southwest side of the hill, most of which support residential patio complexes that are interconnected by stairways and causeways (3).

In addition to the site of La Quemada, the valley contains three major clusters of habitation sites, linked together by an extensive network of raised pathways (4). One cluster, the "La Quemada Aggregate," is located along the banks of the Malpaso River at the base of the hill upon which La Quemada rests. A second, the "Los Pilarillos Aggregate," lies to the southwest of La Quemada and is centered around the small ceremonial site of Los Pilarillos on the Arroyo de Coyotes. The third, and smallest, the "Villanueva Aggregate" is located at the southern end of the valley at the confluence of the Malpaso River and the Arroyo de Coyotes. Each settlement aggregate consists of approximately 50-100 "villages," each composed of residential patio complexes that likely represent households or other social units. Recent excavations indicate that La Quemada and its outlying settlements were initially occupied around AD 500, experienced a period of major architectural growth about AD 700, and were largely abandoned by AD 900 (5).

La Quemada was the central place in the Malpaso settlement hierarchy and most likely was also the center of political power for the immediate region (6). However, imported materials, such as obsidian and marine shell, were not concentrated in the hands of La Quemada residents, but were acquired by all sectors of Malpaso society, both as raw materials and as finished products (7). Further, evidence suggests that craft specialization did not characterize the production industries of chert (8) and obsidian tools (9) or shell ornaments (10). Recent studies suggest that La Quemada's rulers may have exercised their control over subordinate populations in the valley through a combination of ancestor veneration and the institutionalized use of violence, as suggested by the abundance of human skeletal deposits and their associated mortuary patterning focused on above-ground displays of human remains (11).

## Reconstructing Incised-Engraved Ware Technology

Incised-engraved wares were produced in the Malpaso Valley perhaps from AD 500 until approximately AD 900. Incising and engraving occur on highly burnished tripod vessels, fired dark brown to black in color. Designs are restricted to simple, geometric motifs often filled with red and/or white mineral pigments that appear on a continuous band, usually no more than 5 cm wide, just below the rim. It is conceivable that the use of incised-engraved vessels was restricted to certain segments of Malpaso society or to certain activities, as suggested by the relatively low frequencies of vessels and their somewhat sophisticated production technology. Holien and Pickering (12) place Michilia Red-Filled Engraved pottery of the Alta Vista region to the northwest (ca. AD 750-900) in context with aspects of Chalchihuites ceremonialism, and Kelley (13) makes similar observations for Vesuvio Red-Filled Engraved wares (ca. AD 600-750) recovered from the same region. Given that Malpaso incised-engraved pottery may have been reserved for ritual use, it is important to caution that the present study of its manufacturing technology may not reflect trends in plainware pottery.

Incised-engraved wares from the Malpaso Valley can be divided into three ceramic complexes with possible chronological significance based on their particular design styles (Table I) and on similar distinctions made by Kelley and Kelley (14): the Malpaso (ca. AD 450-600), the La Quemada (ca. AD 600-750), and the Murguía (ca. AD 750-900). While the raw materials and resource zones used by ancient Malpaso potters probably changed little over time, the decorative and firing processes underwent significant changes that resulted in greater control over the manufacturing process.

Table I. Ceramic Inventory for the Malpaso Valley.

| | La Quemada Site | | La Quemada Aggregate | | Los Pilarillos Site | | Los Pilarillos Aggregate | | Other Valley Sites | | |
|---|---|---|---|---|---|---|---|---|---|---|---|
| Pottery Wares | n | % | n | % | n | % | n | % | n | % | Totals |
| Plain | 104,532 | 61 | 9,138 | 5 | 23,169 | 13 | 33,711 | 20 | 1,449 | 1 | 171,999 |
| Slipped | 10,969 | 57 | 1,004 | 5 | 3,680 | 19 | 3,434 | 18 | 178 | 1 | 19,265 |
| Painted | 10,307 | 66 | 837 | 5 | 2,000 | 13 | 2,401 | 15 | 93 | 1 | 15,638 |
| Negative | 1,226 | 91 | 13 | 1 | 101 | 7 | 13 | 1 | 0 | 0 | 1,353 |
| Pseudo-cloisonne | 340 | 78 | 0 | 0 | 77 | 18 | 20 | 5 | 0 | 0 | 437 |
| Incised-engraved | 2,141 | 47 | 256 | 6 | 936 | 21 | 1,162 | 26 | 46 | 1 | 4,541 |
| Historic | 0 | 0 | 0 | 0 | 0 | 0 | 207 | 100 | 0 | 0 | 207 |
| Indeterminate | 2,797 | 57 | 231 | 5 | 343 | 7 | 1,516 | 31 | 1 | 0 | 4,888 |
| Totals | 132,312 | 61 | 11,479 | 5 | 30,306 | 14 | 42,464 | 19 | 1,767 | 1 | 218,328 |

Note: All totals and percentages are calculated by row.

## What raw material sources were utilized?

Generally, Malpaso clays are the products of the decomposition of feldspathic rocks, which contain abundant concentrations of alumina ($Al_2O_3$) and silicate ($SiO_2$) along with varying amounts of potassium (K-feldspars, or orthoclase), and sodium and calcium (Na/Ca-feldspars, or plagioclase). Incised-engraved pottery fabrics are medium-textured pastes with angular grains of quartz, sub-angular grains of plagioclase feldspar, and angular shards of volcanic glass. Accessory aplastics include hematite, limonite, and pyroxene, all appearing in protracted quantities compared to other Malpaso fabric classes (15). In order to determine the range of materials used in the manufacturing process of incised-engraved pottery, we sampled and analyzed clay-rich sediments from some of the geological formations exposed by stream cuts along the Malpaso River and the Arroyo Coyotes in the Malpaso Valley. We collected twenty clay samples of approximately .5 liter each from several points across the valley that represented both visual and stratigraphic differences. We processed the samples by drying them, mixing them with distilled water, forming them into small briquettes, and firing them for one hour in an oxidizing atmosphere at 750°C.

We then analyzed each sample with a scanning-electron microprobe and energy-dispersive x-ray (SEM-EDX) spectrometer (16, 17) to determine the range and proportion of elements specific to each clay type. The principal reason for selecting SEM-EDX spectroscopy is the technique's ability to conduct point analyses, thereby avoiding chemical signatures produced from included, clastic materials that might prevent the identification of discrete chemical groups (18, 19). For the present analysis, a JEOL JSM-840 scanning-electron microscope, configured with backscattered and secondary electron detectors and an energy-dispersive spectrometer, was used at the Center for Solid State Science at Arizona State University. The x-ray detector was mounted at a take-off angle of 40°. All assays were made using an accelerating voltage of 15 keV and a beam current of 10 nA. The electron beam diameter was held constant for each analysis at 2 μm, and x-ray counting time averaged 30 seconds per assay. For each sample, three spots were analyzed from different clay particles and the results were averaged together.

In order to determine whether or not the clays that we collected in the Malpaso Valley were similar to those used by ancient Malpaso potters, we also used the SEM-EDX to characterize the chemistries of 57 incised-engraved sherds (from the Malpaso Complex) collected from the La Quemada (n=24) and Los Pilarillos (n=33) settlement aggregates. The SEM-EDX study enabled us to identify two main pottery groups with distinct chemical compositions using a hierarchical cluster analysis and a principal components analysis conducted on ten element concentrations identified for each clay sample and pottery sherd (Figure 2): aluminum (Al), silicon (Si), potassium (K), calcium (Ca), scandium (Sc), titanium (Ti), vanadium (V), chromium (Cr), manganese (Mn), and iron (Fe). The results of the principal components analysis are statistically significant at the $<.01$ level ($\chi^2$ = 1592.8; $df$ = 45; $p$ = .000). A multivariate analysis of variance (Pillai's Trace, $F$=5.05; $df$=10, 46; $p$=0; *power*=.99) of the chemical data further indicates a significant difference between the two compositional groups. In addition, we used a discriminant analysis to evaluate the probability that a given sample belongs to one of the two chemical groups (6). The results indicate that clays from deposits along the Malpaso River (n=10) and the sherds collected from the area of the La Quemada settlement aggregate (n=24) form one chemical group, which is compositionally distinct from the chemical group formed by clays retrieved from the Arroyo Coyotes (n=10) and sherds collected from the Los Pilarillos aggregate (n=33); the former contains clays rich in Al, while the latter contains Fe-rich and Si-rich clays.

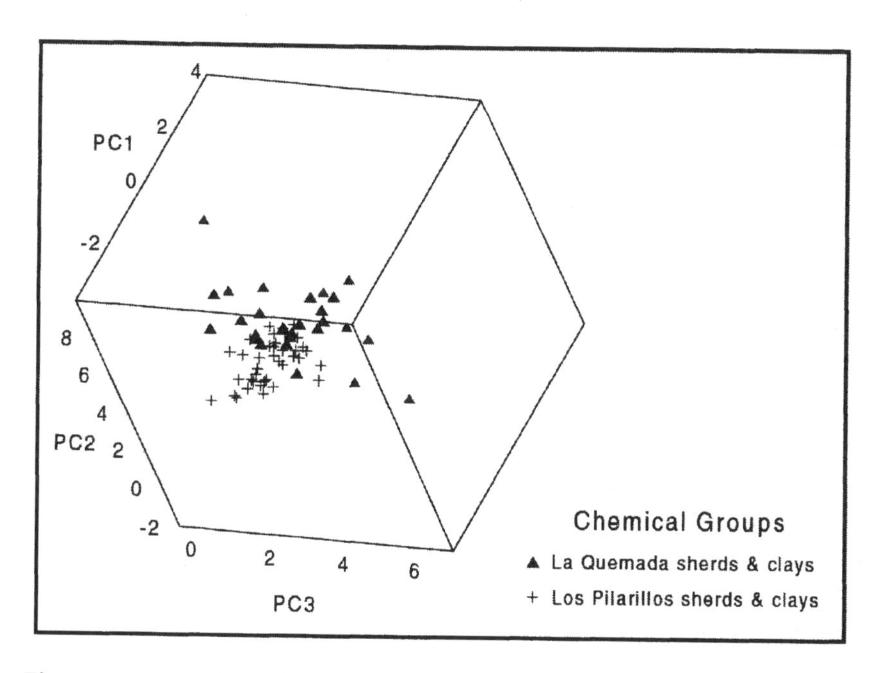


Figure 2. Trivariate plot of the first three principal components (with eigenvectors of the covariance matrix as reference axes) for incised-engraved pottery compositions indicating chemical group affiliations among ceramics and clays.

To corroborate the results of the quantitative analyses, that is, to evaluate whether or not the associations among clays and pottery reflect real compositional groups and not statistical inventions, we re-fired 30 of the pottery sherds at 900°C for three hours in an oxidizing kiln (15), a method that is used often in ceramic studies to provide visual distinctions between pottery from different resource zones (20, 21). Re-fired pottery sherds from the La Quemada settlement aggregate (n=15) tended to turn white (Munsell value, 10YR 8/1) to pink (7.5YR 7/4) in color, while re-fired sherds from the Los Pilarillos aggregate (n=15) tended to turn light red (2.5YR 6/6) to red (2.5YR 5/6). These color changes, most likely the result of different concentrations of Fe in the samples (higher concentrations lead to darker hues of red when re-fired; 22) support the differences indicated by the chemical analyses.

Together, these analyses indicate that Malpaso potters utilized at least two chemically distinct raw material sources, although the degree to which the sources correspond to the clays situated in the valley's settlement aggregates needs to be evaluated further with petrographic analyses and additional chemical assays. If there is a correspondence between raw material sources and settlement aggregates, then it can be suggested that pottery manufacture likely was not highly specialized during the earliest occupation of the valley (ca. AD 450-600), since multiple production centers located in at least two separate settlement clusters would not tend to yield standardized products (23).

## How were the vessels formed and decorated?

To reconstruct the forming process, we analyzed five partially complete vessels from all ceramic complexes (Malpaso, n=1 from Los Pilarillos; La Quemada, n=2 from Los Pilarillos; Murguía, n=2 from La Quemada). Macroscopic examination indicates that the sequence began with primary forming, probably by a coiling method as evidenced by a set of parallel coils found running horizontally near the base of one bowl. Secondary forming most likely involved a combination of scraping and beating, such as with the use of a paddle-and-anvil. Evidence consists of visible scraping marks on the interior surface of one bowl, and a series of slight depressions in the interior wall of a different vessel that suggests the use of an anvil to support blows to the vessel exterior. The final forming step involved the addition of three supports to a vessel bottom, which also appear to have been scraped smooth.

To analyze the decorating process, we examined 620 sherds, collected from settlement survey in the valley, for pigment use (red or white) and line type, that is, whether the sherd exhibited incised or engraved lines (24), or a combination of the two (Figure 3). Decorations on pottery from the Malpaso Complex (n=427), the earliest of the three ceramic complexes in the valley, were either completely incised, completely engraved, or involved a combination of the two; incised lines, however, were used to decorate the majority of sherds we examined from this complex. Designs were incised into the vessels when leather-hard, after they had been lightly burnished or in some cases polished, and red and white pigments were rubbed into the design zone after firing. Engraved lines, on the other hand, were made at the end of the manufacturing process, after the vessel had been fired, along with the application of red and white pigments. The designs depicted on the majority of pots from the La Quemada Complex (n=125), which followed, were most often completely engraved, although a few sherds also had incised lines. The decorations on Murguía pottery (n=68), believed to be the latest of the ceramic complexes, were completely engraved, with no evidence of incising. In fact, in most cases the engraving is so heavy that it almost replicates a champlevé technique, in which large portions of the vessel are carved out and filled with pigment. Pigment use also seems to change during this time. Instead of using both red and white in alternating patterns as with the other complexes, potters used either solely red or solely white to decorate Murguía vessels.

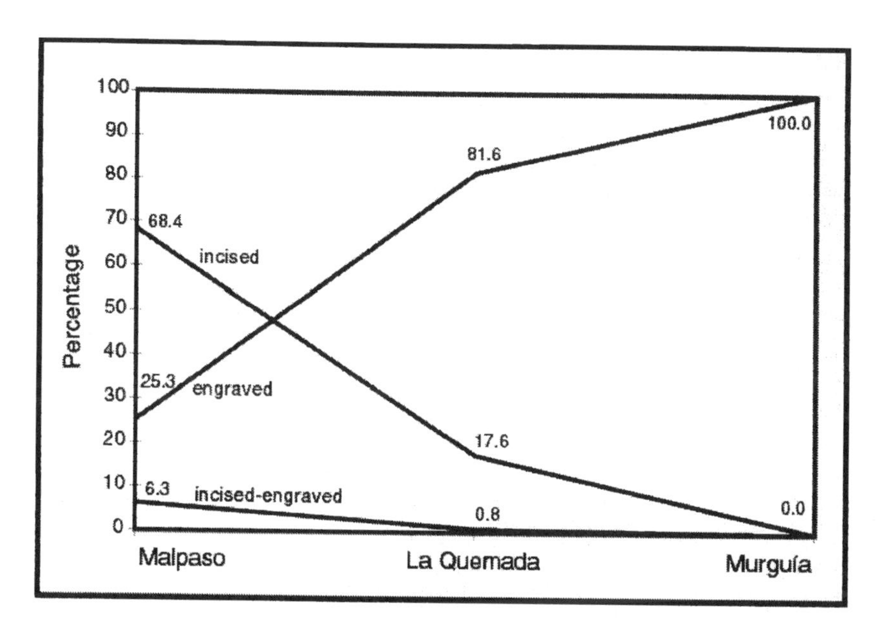


Figure 3. Line graph of the frequencies of incised, engraved, and incised-engraved ceramics in the Malpaso Valley through time. Point 1 on the x-axis corresponds to the period (ca. A.D. 450-600) represented by Malpaso Complex incised-engraved ceramics (n=427), point 2 corresponds to the period (c. A.D. 600-750) represented by the La Quemada Complex incised-engraved ceramics (n=125), and point 3 corresponds to the period (c. A.D. 750-900) represented by the Murguía Complex incised-engraved ceramics (n=68).

**What techniques were involved in the firing process?**

Two different methods are commonly used in the determination of ancient firing temperatures and the performance of the firing apparatus. In the first, a rough indication of the firing temperature is obtained through studying the minerals present in the ceramic by means of X-ray diffraction techniques, optical spectroscopy, and differential thermal analysis (25). In the second, the firing temperature is estimated using the data obtained from thermal expansion measurements on the ceramic (26).

In the present study, we used a third technique based on previous work by other archaeologists (20, 21, 24) in which sherds are subject to a kiln re-firing process that allows for the determination of a temperature range over which the pottery was originally fired. We selected 100 sherds at random from all three ceramic complexes that were surface collected from valley settlement surveys (Malpaso, n=40; La Quemada, n=35, Murguía, n=25), and removed a small chip from each one. We noted the original state of oxidation and color of the chips using a Munsell color chart, and then observed the rate of color change as the chips were heated through progressively higher temperatures. The point at which the color changes as a result of the re-firing usually indicates the upper end of the interval closest to the original firing temperature. We began firing the chips at 500°C and re-fired them at 100°C intervals until 1000°C, firing 5 minutes for each interval in an oxidizing atmosphere. At each interval we recorded the color changes in the core and surface of each chip. To refine the observations made during this analysis, we fired the raw clay samples from the clay sourcing survey at 10°C intervals between 600°C - 900°C for 3 hours, and compared their colors to the original colors of the sherds used in the re-firing study.

The results indicate that there is great variability in the firing temperatures of sherds from the Malpaso Complex (n=38), from 600°C-900°C (Figure 4). The firing range appears to have been shorter during the La Quemada Complex (n=34), from 700°C-800°C. Murguía Complex (n=28) sherds all seem to have been fired between a much tighter range, from 760°C-780°C. While no evidence of ceramic combustion devices have been recovered from the region, these relatively low firing temperatures could have been achieved using simple open pit firing, although the relatively narrow temperature range achieved for firing Murguía pottery suggests a more sophisticated pyrotechnology, such as the use of a formal kiln.

## Discussion

The results of these analyses allow for the reconstruction of the production process, and also suggest that certain steps in the decorating and firing processes were manipulated by local producers, which resulted in significant technological changes. Importantly, the data document a change in the style of line execution over time. The majority of (probably early) Malpaso wares were incised before being fired, although a few examples show both incising and engraving. Most of the (probably later) La Quemada wares, on the other hand, were completely engraved after firing had taken place, as were all of the (probably latest) Murguía examples. This pattern indicates a fundamental

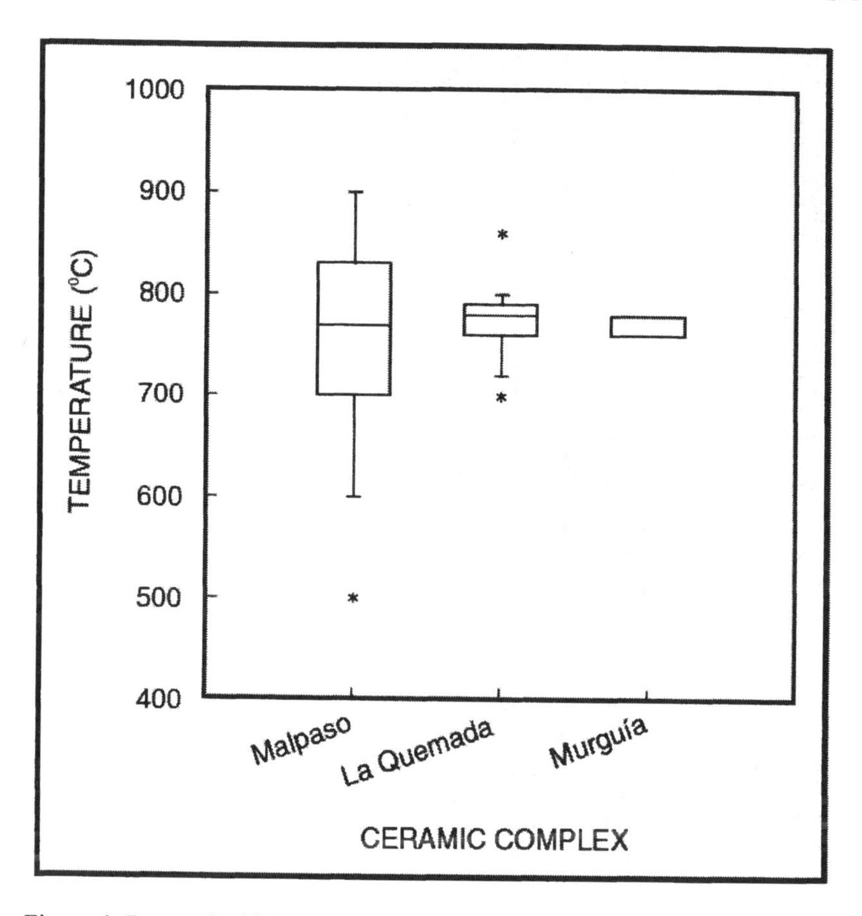


Figure 4. Box-and-whisker plot of the results of the re-firing experiment. The x-axis shows the incised-engraved ceramic complexes, arranged from early (left) to late (right): Malpaso (n=38), La Quemada (n=34), Murguía (n=28).

technological change. By engraving decorations in the pottery after firing rather than incising decorations before firing, the decoration process is moved to the end of the production sequence. This has two important consequences. First, the potential now exists for decoration to be separated from the rest of the production process, thereby allowing for the participation of individuals who specialize in vessel decoration (23). Second, this change in production steps increases the efficiency of the manufacturing process. Potters no longer have to expend time and energy decorating vessels that eventually may be destroyed in the drying and firing processes. Post-fire decoration ensures that the time and energy invested in the decoration of the vessel will not go wasted. In addition, the shift from using both red and white pigments in decoration to only one pigment during the Murguía phase also likely resulted in increased efficiency. In addition, these patterns suggest that the changes may have been gradual over the period of AD 600-750 and may have included several different potters, since the La Quemada wares evince both incising and engraving at this time.

Finally, the wide range of firing temperatures for Malpaso wares suggests that early potters did not have complete control over the firing process, or that they did not exercise that control. The firing temperature range for La Quemada wares is noticeably shorter than that of Malpaso wares, while the range for Murguía wares appears to have been very tight, only twenty degrees. The establishment of a consistent firing temperature parallels the shift seen in the decorative process where later ceramics were fashioned in a more standardized way compared to earlier ceramics. It appears that by the time of Murguía ware production, late in the ceramic sequence, potters had refined both the firing and decorating processes.

These findings have important implications for understanding political and economic organization in the Malpaso Valley. For example, increases in the specialization of incised-engraved wares over time could indicate a greater degree of administration in their production and distribution by certain segments of society (27). If post-fire engraving and standardized firing methods involved specialized knowledge and technical skills, then this could have led to the development of certain social entities, such as specific households or potting groups, devoted to the specialized task of pottery manufacture (28). In addition, if incised-engraved wares were restricted to ritual use in the Malpaso Valley, as has been suggested for similar wares in adjacent regions (12, 13), the changes in technological attributes recorded in this study could reflect increasing control over ritual behavior (or at least the material components of certain rituals) versus changes in political or economic organization.

The present study demonstrates that Malpaso potters changed some of their techniques over time so as to create some of the conditions conducive to craft specialization. These technological innovations could have involved the division of operational tasks that allowed different individuals to devote their energies to a single, specific task, such as firing or decoration (29, 30). Continued studies of the technology involved in the manufacture of incised-engraved pottery, combined with a clearer understanding of their distributions over time, will allow us to evaluate these implications more thoroughly, as well as to model the effects of ceramic production, distribution, and consumption on the sociocultural development of the Epiclassic Malpaso Valley.

*Acknowledgements*. The work reported here is part of the La Quemada-Malpaso Valley Archaeological Project and was conducted with the permission of the Consejo Nacional de Arqueología of the Instituto Nacional de Antropología e Historia, Mexico. We gratefully acknowledge the aid of Ing. Joaquín García Bárcena, Dr. Alejandro Martínez Muriel, and Dr. José Francisco Román Gutierrez of that institution. The support and advice of Peter Jiménez Betts have been important throughout the project. Samples were obtained through grants from the National Science Foundation (Grant Nos. BNS-8806238 and DBS-9211681), the National Endowment for the Humanities, the Wenner-Gren Foundation for Anthropological Research, and the Foundation for the Advancement of Mesoamerican Studies. Microprobe analysis was conducted at the Center for Solid State Science at Arizona State University (ASU) with the help of Robert Roberts and funding by a grant from the Research and Development Committee, Department of Anthropology, ASU. Data processing was carried out at the Archaeological Research Institute at ASU with the support of Arleyn Simon and funds from Sigma Xi, The Scientific Research Society. We would like to thank Karla Davis-Salazar, Kathryn Jakes, and an anonymous reviewer for their numerous helpful comments on drafts of this work that greatly improved its clarity. In addition, Margaret O'Day, Vincent Schiavitti, Thomas Sharp, Arleyn Simon, Nicola Strazicich, and Paula Turkon are thanked for their intellectual and editorial contributions to previous versions of this study.

## References

1. Strazicich, N. M. *Latin American Antiquity*. **1998**, *9*, 259-274.
2. Nelson, B. A. In *Culture and Contact: Charles C. DiPeso's Gran Chichimec*; Wooseley, A.; Ravesloot, J., Eds.; University of New Mexico Press, Albuquerque, 1994; pp. 173-190.
3. Nelson, B. A. *American Antiquity*. **1995**, *60*, 597-618.
4. Trombold, C. D. In *Ancient Road Networks and Settlement Hierarchies in the New World*; Trombold, C., Ed.; Cambridge University Press: Cambridge, England, 1991; pp. 145-168.
5. Nelson, B. A. *Journal of Field Archaeology*. **1997**, *24*, 85-109.
6. Wells, E. C. *Latin American Antiquity*. **2000**, *11*, 21-42.
7. Wells, E. C.; Nelson, B. A. In *Producción e intercambio de recursos estratégicos en el occidente de México*; Williams, E., Ed.; Colegio de Michoacán: Zamora, Michoacán, México, 2002.
8. Kantor, L. M. M.A. thesis, State University of New York, Buffalo, NY, 1995.
9. Millhauser, J. K. M.A. thesis, Arizona State University, Tempe, AZ.

10. Vargas, V. D. La Quemada-Malpaso Valley Archaeological Project Shell Artifacts. Unpublished manuscript, Arizona State University, Tempe, AZ.
11. Nelson, B. A.; Darling, J. A.; Kice, D. A. *Latin American Antiquity*. **1992**, 3, 298-315.
12. Holien, T.; Pickering, R. B. In *Middle Classic Mesoamerica: A.D. 400-700*; Pasztory, E., Ed.; Columbia University Press: New York, 1978.
13. Kelley, J. C. In *Archaeology of Northern Mesoamerica*; Ekholm, G.; Bernal, I., Eds.; Handbook of Middle American Indians; University of Texas Press: Austin, TX, 1971; Vol. 11, pp. 780-797.
14. Kelley, J. C.; Kelley, E. A. *An Introduction to the Ceramics of the Chalchihuites Culture of Zacatecas and Durango, Mexico, Part I: The Decorated Wares*; Mesoamerican Studies 5; University Museum, Southern Illinois University: Carbondale. IL, 1971.
15. Wells, E. C.; Nelson, B. A. In *Estudios cerámicos en el occidente y norte de México*; Williams, E.; Weigand, P., Eds.; Colegio de Michoacán: Zamora, Michoacán, México, 2001.
16. Reed, S. J. B. *Electron Microprobe Analysis*; Cambridge University Press: Cambridge, England, 1975.
17. DeAtley, S. P.; Blackman, M. J.; Olin, J. S. In *Archaeological Ceramics*; Olin, J.; Franklin, A., Eds.; Smithsonian Institution Press: Washington, DC, 1982; pp. 79-88.
18. Freestone, I. C. *Archaeometry*. **1982**, *24*, 99-116.
19. Tite, M. S.; Freestone, I. C.; Meeks, N. D.; Bimson, M. In *Archaeological Ceramics*; Olin, J.; Franklin, A., Eds.; Smithsonian Institution Press: Washington, DC, 1982; pp. 109-120.
20. Rice, P. M. *Pottery Analysis: A Sourcebook*; University of Chicago Press: Chicago, IL, 1987; pp. 343-345, 427-429.
21. Shepard, A. O. *Ceramics for the Archaeologist*; Carnegie Institution of Washington: Washington, DC, 1976; pp. 104-107.
22. Hess, J.; Perlman, I. *Archaeometry*. **1974**, *16*, 137-152.
23. Costin, C. L.; Hagstrum, M. B. *American Antiquity*. **1995**, *60*, 619-639.
24. Rye, O. S. *Pottery Technology: Principles and Reconstruction*; Taraxacum: Washington, DC, 1981; pp. 90-91.
25. Roberts, G. *Archaeometry*. **1960**, *3*, 36-37.
26. Tite, M. S. *Archaeometry*. **1969**, *11*, 131-144.
27. Brumfiel, E. M.; Earle, T. K. In *Specialization, Exchange, and Complex Societies*; Brumfiel, E.; Earle, T., Eds.; Cambridge University Press, Cambridge: England, 1987; pp. 1-9.
28. Stark, M. *World Archaeology*. **1991**, *23*, 64-78.
29. Costin, C. L. In *Archaeological Method and Theory*; Schiffer, M., Ed.; University of Arizona Press: Tucson, AZ, 1991; Vol. 3, pp. 1-56.
30. Hagstrum, M. B. *Journal of Field Archaeology*. **1985**, *12*, 65-75.

## Chapter 13

# Trace Element Analysis and Its Role in Analyzing Ceramics in the Eastern Woodlands

**Christina B. Rieth**

**New York State Museum, Division of Research and Collections, Cultural Education Center 3122, Empire State Plaza, Albany, NY 12230**

Clemson Island and Owasco ceramics are used to document interaction in Pennsylvania during the Early Late Prehistoric Period (A.D. 700-1300). Clemson Island ceramics are regarded as the by-products of local manufacture while Owasco ceramics are considered to be foreign wares. X-ray fluorescence is used to determine the provenance of manufacture and assess whether compositional profiles reflect the reciprocal exchange of ceramic vessels. The results of this project suggest no clear distinction between ceramic types and clay deposits. Instead, similarities in attributes may reflect the utilization of similar resources.

Ceramic vessels represent important units of analysis in Early Late Prehistoric (A.D. 700-1300) interaction studies. In north-central Pennsylvania, the Early Late Prehistoric Period is characterized by the Clemson Island ceramic tradition whose spatial distribution is concentrated along the main and west branch of the Susquehanna River (*1,2*). Throughout this region, ceramics belonging to the northerly Owasco tradition have been recovered and are often regarded as the by-products of long-distance interaction between groups. More recently, however, archaeologists have questioned this assumption suggesting that the relationship between these populations may be more complex with Clemson Island and Owasco groups occupying a similar ecological niche (*3,4*).

 

If the location of manufacture of these vessels can be determined, such information would not only contribute to our understanding of the use of local resources but would also provide information about the settlement patterns of these prehistoric populations.

The chemical characterization of ceramic pastes using trace element analysis provides a viable mechanism for elucidating such information. In this chapter, I discuss the preliminary results of a project designed to assess the chemical composition of ceramic sherds from four Early Late Prehistoric sites in north-central Pennsylvania. This research was designed to address questions related (1) to the provenance of manufacture of Owasco vessels, and (2) to determine whether there is any evidence for long distance trade of Owasco vessels at Clemson Island sites. When combined with information from local clay deposits, it is suggested that there is no clear distinction between ceramic types and the clay deposits exploited by these two groups. Instead, similarities in the technological attributes of these containers may more accurately reflect the utilization of similar resource procurement zones.

## Background

Sites associated with the Clemson Island tradition date between A.D. 700 and 1300 and are largely found along the north and west branches of the Susquehanna River. Owasco sites generally date between A.D. 900 and 1300 and are concentrated in central and southern New York. The spatial distribution of Owasco sites minimally overlaps the Clemson Island tradition with a few camps reported in north-central Pennsylvania. The settlement patterns of these groups are oriented around agriculturally based hamlets and smaller resource procurement sites.

Ceramic sherds are regularly used to distinguish between Clemson Island and Owasco site. Clemson Island and Owasco sherds are very similar and both exhibit cordmarked motifs, outflaring rim, grit temper, and flat lip shape. Hay et al. (*5*) indicate that some early Clemson Island ceramics can be identified by the presence of punctates and overstamping around the rim. Archaeologists also argue that Clemson Island vessels can be distinguished by their ceramic paste. According to Smith (*6*), "Owasco ceramics have a denser and more integrated paste than their Clemson Island look-alikes, which are often poorly fired and appear laminated in cross section".

The presence of Owasco sherds on Clemson Island sites is often interpreted as evidence of trade between the two groups (*4*). However, archaeologists have recently questioned this assumption suggesting that trade alone could not account for the large number of Owasco sherds recovered from Clemson Island sites (*1*). Instead, archaeologists argue that other mechanisms (including inter-group marriage and migration) may more accurately explain the presence of

Owasco vessels on Clemson Island sites (*1*). Recently, archaeologists have also suggested that Owasco and Clemson Island ceramics may not represent different types but may represent local variations on a larger regional type (*7, 8*). Although studies centering on the provenance of ceramic vessels have not been completed in the past, Stewart (*1*) argues that, if completed, such studies could provide important information that could be used to assess the validity of these hypotheses.

In any study of the provenance of ceramic clays it is important to understand that the selection of clays by prehistoric potters was not haphazard but probably represented a balance between technical and nontechnical choices. Quality is one of the leading factors in the selection of clays. In their study of Kalinga ceramics, Aronson et al. (*9*) indicate that potters preferred clays with few mineral inclusions and a high degree of plasticity. Accessibility and distance were also important factors in selection; potters preferred to procure materials from deposits near their house or within a a familiar political region. This coincides with Arnold's (*10*) estimates that Native populations procured clays from deposits located less than 50 km from their primary residence.

The remainder of this chapter discusses the results of a project designed to assess the chemical composition of ceramic sherds from four sites located in north-central Pennsylvania. The goal of this research was to determine the provenance of manufacture of Owasco vessels, and assess evidence for the trade of Owasco vessels at Clemson Island sites.

## Overview of Sites and Ceramic Assemblage

Sherds from the Tioga Point Farm, Wells, St. Anthony's, and Fisher Farm sites form the basis for this study. Tioga Point Farm (*11, 12*) and Wells (*13*) are camps located near the New York-Pennsylvania border and are located in an ambiguous area occupied by both Clemson Island and Owasco groups (Figure 1). St. Anthony's (*1*) and Fisher Farm (*2*) are agricultural hamlets located in central Pennsylvania. Both sites are located in "the Clemson Island heartland" and are believed to have been occupied by Clemson Island groups. Although Fisher Farm and Tioga Point Farm were repeatedly occupied, the sherds used in this project were retrieved from occupations dating to the Early Late Prehistoric Period (c. A.D. 900 to 1300).

The ceramics from these sites exhibit exterior cordwrapped stick and paddle impressions around the rim and neck (*14*). The interior rim and lip of these vessels were also decorated with cordmarked motifs. Clemson Island sherds were distinguished from their Owasco counterparts by the presence of presence of exterior punctates and interior beads around the rim (*1,2*). Most sherds were grit tempered with quartz, quartzite, and gneiss being major constituents (*1, 13*).

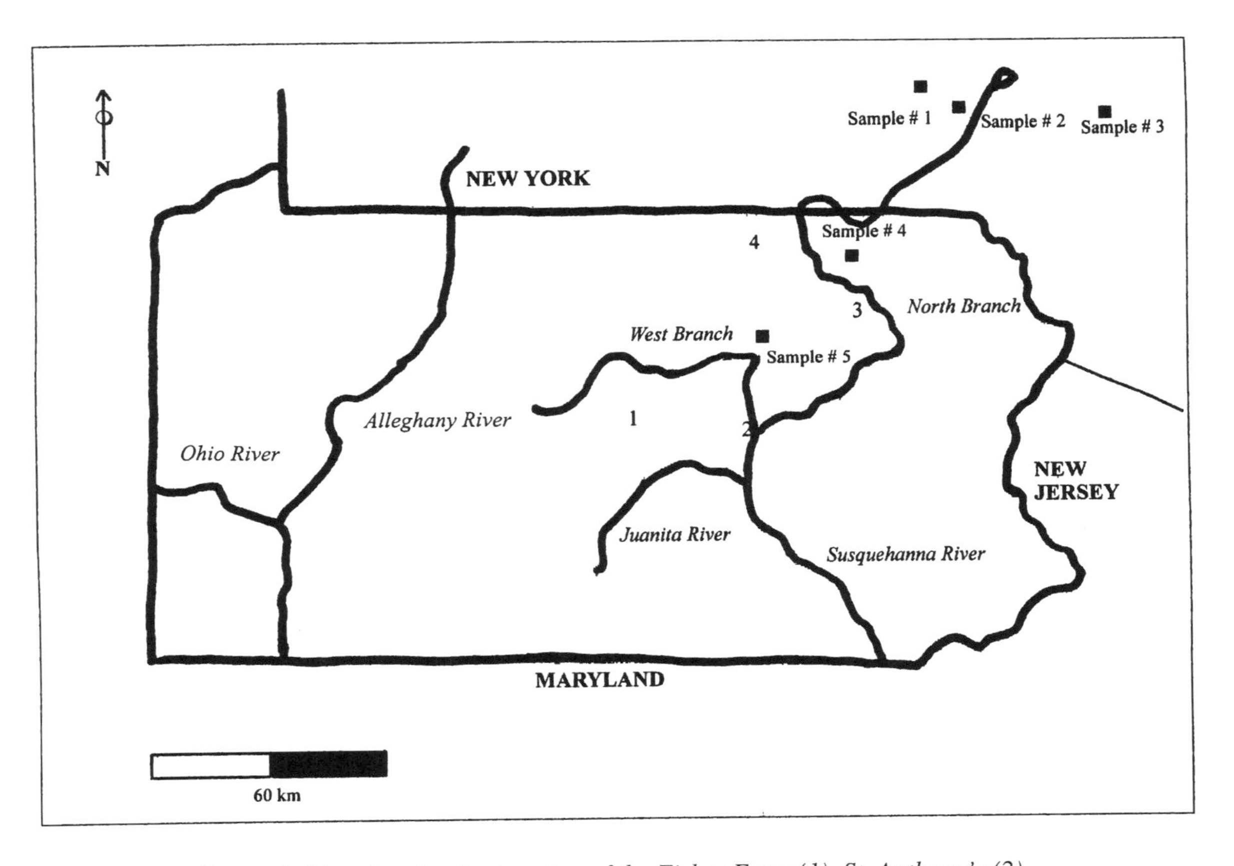


*Figure 1. Map showing the location of the Fisher Farm (1), St. Anthony's (2), Wells (3), and Tioga Point Farm (4) sites. Clay samples 1-5 are also shown.*

Chert, shell, and limestone tempered sherds were also identified. The size of the inclusions averaged 1-2 mm and were poor to moderately well sorted.

## Methodology

### Sample Preparation

Forty-six sherds were analyzed from these four sites (Table I). Each sherd was determined to have originated from a distinct vessel lot based on technological and decorative attributes. Ceramic sherds derived from containers associated with both traditions. Owasco and Clemson Island sherds were selected from the same radiocarbon dated features to insure against mixing of sherds from different occupations. Sample preparation included brushing and washing the surface of the artifact.

Data were also collected on five raw clay samples from the Susquehanna Valley (Figure 1). The clay samples collected have not been extensively surveyed but represent samples recovered during archaeological and geological survey. Using a technique similar to that described by Neff et al. (*15*), approximately 2 kg of clay were collected from secondary deposits located along stream cuts of major and minor tributaries of the Susquehanna River. Each sample was dried and crushed before water was added. The resulting matrix was formed into a small tile measuring 3 cm in diameter and was fired in an oxidizing atmosphere at 850°C. When used in conjunction with data from ceramic sherds, the compositional profiles of these clay samples provide a basis for determining whether ceramic containers were locally manufactured.

**Table I. Total Number of Sherds Analyzed**

| *Ceramic Type* | *Tioga Point Farm* | *Wells* | *St. Anthony's* | *Fisher Farm* | *Total* |
|---|---|---|---|---|---|
| Owasco | 10(21.7%) | 3(6.5%) | 3(6.5%) | 3(6.5%) | 19(41.3%) |
| Clemson Is. | 4 (8.7%) | 6(13%) | 6(13%) | 11(23.9%) | 27(58.2%) |
| Total | 14(30.4%) | 9(19.6%) | 9(19.6%) | 14(30.4%) | 46(100%) |

## Chemical Analysis

All samples were analyzed using x-ray fluorescence (XRF) at the Accelerator Laboratory of the University at Albany, SUNY. XRF was chosen because of its availability to the researcher and its relatively non-destructive nature allowing museum collections to be analyzed. Since these artifacts represent irreplaceable artifacts, portions of the matrix were not removed for analysis as has been the practice in similar studies (*16, 17*). Instead, all readings were taken from the exterior surface of the artifact. Two readings were taken from each sherd to insure homogeneity within a single vessel (*18*).

Many researchers have pointed out that the chemical composition of the sherd can be affected by the type and number of inclusions in the paste (*17, 19, 20*). Although Owasco and Clemson Island sherds generally lack large inclusions on the exterior surface, an attempt was made to avoid such inclusions when present by focusing the x-ray's beam on the clay matrix. This approach has also been used by Kuhn and Sempowski (*20*) who argue that since such a small area is measured, the likelihood that the final results would be biased is limited. Furthermore, since most of the sherds in this study exhibited a similar grit temper, biases stemming from different inclusions are expected to be limited. Food encrustations, leaching, and other surface materials can also affect the final compositional profile (*20*). Sherds exhibiting these characteristics were avoided during this study.

The technique used in this study allowed proporational data about the relationship of the elements to be collected. As discussed in Kuhn (*21*), proportional data allow ratios of trace elements to be measured in terms of "the number of characteristic x-rays observed in fixed time". Following bombardment of the sample with a Cadmium 109 radioisotope for one hour, fluorescent x-rays were emitted whose energies are characteristic of the elements present in the sherd (Figure 2). The energy levels of each element were recorded numerically allowing concentrations of specific elements to be determined. Sixteen elements were measured including rubidium (Rb), strontium (Sr), iron (Fe), lead (Pb), vanadium (V), potassium (K), zinc (Zn), nickel (Ni), barium (Ba), yttrium (Y), titanium (Ti), scandium (Sc), manganese (Mg), copper (Cu), and zirconium (Zr). Vanadium (V) and potassium (K) were removed from the final analysis due to low detection limits. Although the number and types of elements used in this study are sufficient for distinguishing between regional clay profiles, in most cases, additional samples and elements are needed to identify the precise location of a specific clay deposit. Since this was beyond the scope of this project, the results are only discussed in terms of the regional clay profiles.

The recorded energy levels of each sherd were collected and stored for analysis using the computer program AXIL (*22*). AXIL was applied to measure peak area counts and standard deviations for each element. Differences in peak counts were recorded in tabular and graphic formats (Figure 2). Multivariate

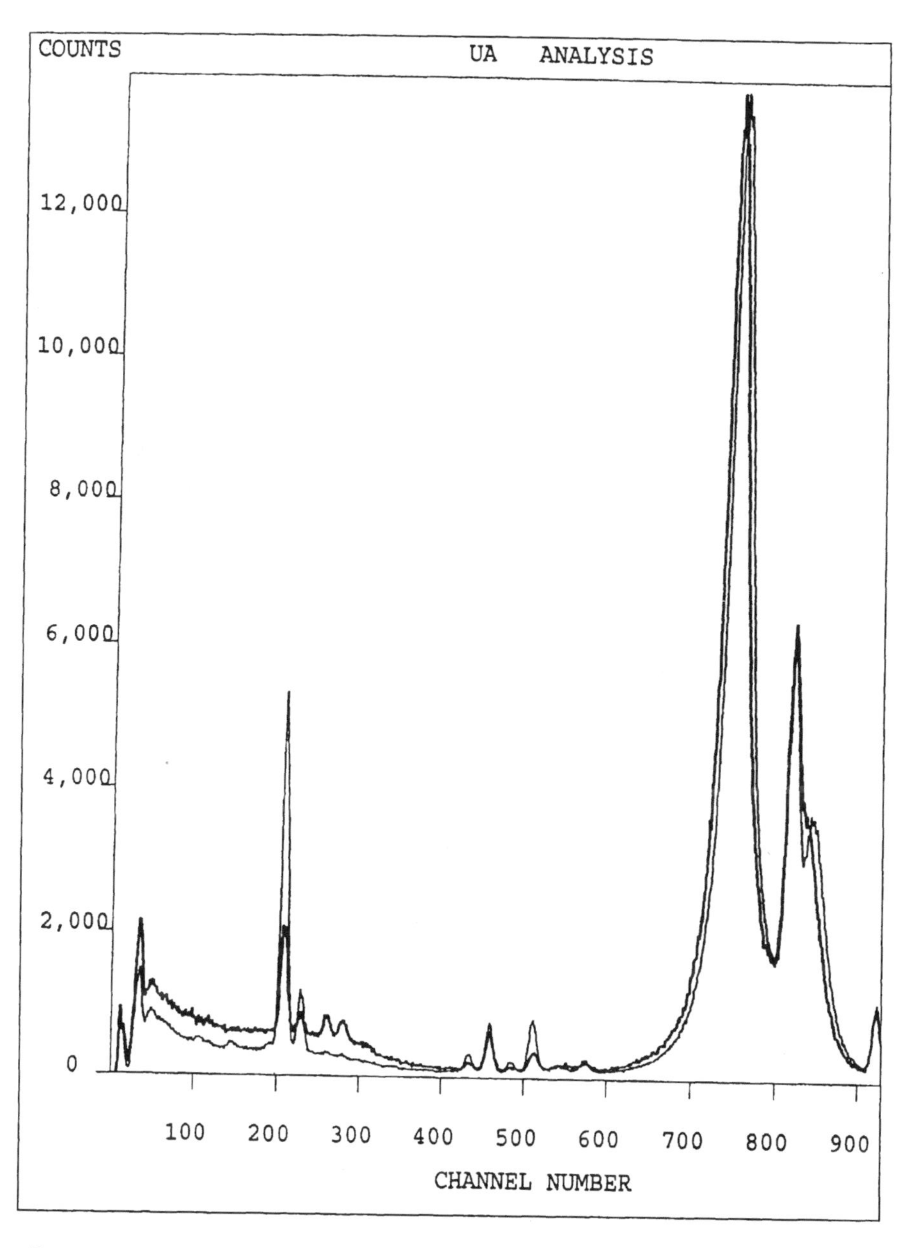


*Figure 2. XRF spectra showing channel counts for sherd recovered from Tioga Point Farm.*

statistics, including principal components analysis, discriminant function analysis, and cluster analysis, were also applied to assess the degree of homogeneity between data. Multivariate statistics were performed using SPSS/PC+ (Version 4.0) (*23*).

## Results

Five clay groups were identified and are here referred to as Otsego, Upland, Tioga, West Branch, and Unknown (Tables II and III). Principal components analysis showed that 91.9% of the total variance in the data set could be explained by the first four components (Table IV). Figure 3 illustrates the spatial relationship of the clay groups to each other.

The Tioga group is the largest group with 30 members from all four sites (Tables II and III). Clemson Island and Owasco sherds both grouped here. Also in this group is Clay sample 4, which was extracted from a deposit in Bradford County, Pennsylvania. This group includes artifacts from features dating from the tenth to the thirteenth century suggesting continuity in the selection of regional deposits during this 300-year period.

The loosely clustered members of the Tioga group suggest that several different deposits may have been used by the occupants of north-central Pennsylvania (Figure 3). The period A.D. 1000 to 1300 is often characterized by increased sedentism, territoriality, and warfare (*1*). If this were true, we should expect tighter clusters suggesting that the Owasco and Clemson Island occupants of north-central Pennsylvania may have continued to practice a semi-sedentary settlement pattern during this time period. Alternatively, warfare may not have been excessive, allowing prehistoric groups to move across clan and tribal territories to gather needed resources (*14*).

**Table II. Grouping of Owasco Sherds**

| *Clay Group* | *Tioga Point Farm* | *Wells* | *St. Anthony's* | *Fisher Farm* | *Clay Samples (Sample No.)* | *Total* |
|---|---|---|---|---|---|---|
| Otsego | 3 | --- | --- | 1 | 2(1,2) | 6 |
| Upland | --- | --- | --- | --- | 1(3) | 1 |
| Tioga | 5 | 3 | 1 | 2 | 1(4) | 12 |
| West Br. | --- | --- | --- | --- | 1(5) | 1 |
| Unknown | 2 | --- | 2 | --- | --- | 4 |
| Total | 10 | 3 | 3 | 3 | 5 | 24 |

**Table III. Grouping of Clemson Island Sherds**

| *Clay Group* | *Tioga Point Farm* | *Wells* | *St. Anthony's* | *Fisher Farm* | *Clay Samples (Sample No.)* | *Total* |
|---|---|---|---|---|---|---|
| Otsego | --- | --- | --- | --- | 2(1,2) | 2 |
| Upland | --- | --- | --- | --- | 1(3) | 1 |
| Tioga | 2 | 6 | 1 | 9 | 1(4) | 19 |
| West Br. | --- | --- | 2 | 2 | 1(5) | 5 |
| Unknown | 2 | --- | 3 | --- | --- | 5 |
| Total | 4 | 6 | 6 | 11 | 5 | 32 |

The large number of sherds from Fisher Farm is interesting and provides some evidence for the non-local production of both Owasco and Clemson Island ceramics. As shown in Figure 1, the Fisher Farm site is located approximately 80 km from Clay sample 4. The geographic distance between Clay sample 4 and the Fisher Farm site exceeds the distance that most groups would have been expected to travel to procure clays suggesting that a more extensive reciprocal exchange network may have existed between the occupants of northern and central Pennsylvania.

The Otsego group produced 4 sherds with 1 Owasco sherd from Fisher Farm and 3 Owasco sherds from Tioga Point Farm. Clay samples 1 and 2, which were collected from deposits in Otsego County, New York, also clustered in this group. The geographic distance between Clay samples 1 and 2 and Tioga Point

**Table IV. Principal Components Analysis for Samples**

| *Principal Component* | *Eigenvalue* | *Percent Variation* | *Cumulative Percentage* |
|---|---|---|---|
| 1 | 3.962 | 44.0 | 44.0 |
| 2 | 2.266 | 25.2 | 69.2 |
| 3 | 1.098 | 12.2 | 81.4 |
| 4 | 0.9418 | 10.5 | 91.9 |
| 5 | 0.4049 | 4.5 | 96.4 |
| 6 | 0.2155 | 2.4 | 98.8 |
| 7 | 0.0565 | 0.6 | 99.4 |
| 8 | 0.0346 | 0.4 | 99.8 |
| 9 | 0.0184 | 0.2 | 100.0 |

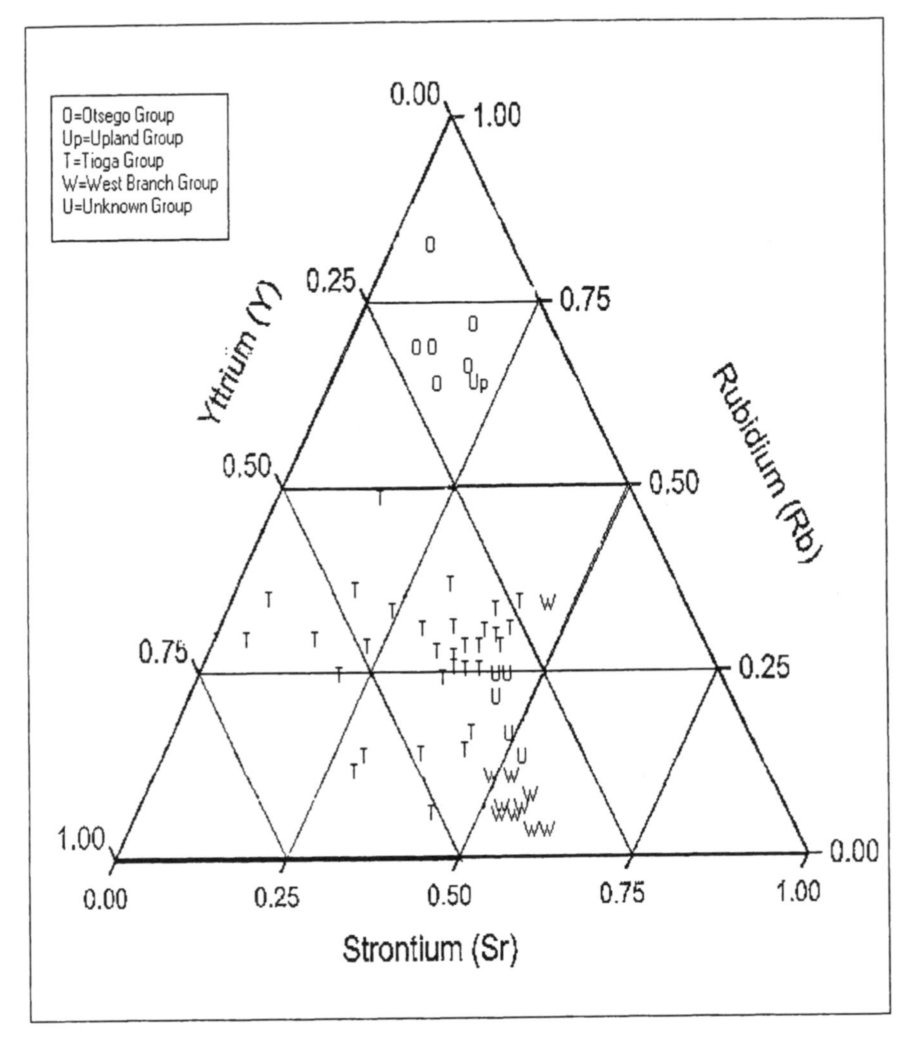


*Figure 3. Spatial distribution of clay groups and clay samples.*

Farm is 80 km while the distance between these clay samples and Fisher Farm is approximately 120 km. Given the large distance between the sites and clay samples 1 and 2, it seems likely that these artifacts also represent the by-products of reciprocal exchange.

Two Clemson Island sherds from St. Anthony's, 2 Clemson Island sherds from Fisher Farm, and Clay sample 5 form the West Branch Group (Table III).

St. Anthony's, Fisher Farm, and clay sample 5 are located in central Pennsylvania south of the West Branch of the Susquehanna River. This area is the "heartland" of the Clemson Island tradition and it seems reasonable that only Clemson Island sherds grouped here.

Nine sherds comprise the Unknown group with 4 from Tioga Point Farm and 5 from St. Anthony's. Five sherds were assigned to Clemson Island types and 4 to Owasco types (Tables II and III). None of the clay samples clustered in this group making it difficult to determine the provenance of manufacture. The Unknown group represents a tight cluster of data points that are spatially related to the Tioga and West Branch Group suggesting that the source of the clay may be located north of the West Branch of the Susquehanna River (Figure 3).

The Upland group consists only of Clay sample 3 (Tables II and III). This sample, which was recovered from a deposit located at the northeast end of the Susquehanna Valley, is most closely related to the Otsego Group (Figure 3). Given that only one member grouped in this category it is difficult to say anything conclusive about the relationship between the manufacture of pottery and ceramic typology. However, the failure of any sherds to cluster in this group suggests either that the occupants of north-central Pennsylvania did not maintain interaction networks with groups residing in this area or that they did not use that particular clay deposit (*14*).

In summary, this study suggests that ceramic vessels from the four sites can be grouped into five clay groups: Tioga, West Branch, Otsego, Upland, and Unknown. The members of these groups do not correspond with a single ceramic
tradition but include vessels belonging to both the Owasco and Clemson Island traditions. Grouping of Owasco and Clemson Island containers from the same features within the same group suggests that the manufactures of these vessels exploited similar deposits. Correlation of compositional data from these vessels with similar data from known clay deposits suggests that most of these vessels are not the by-products of reciprocal exchange but rather represent locally manufactured containers.

## Discussion

Trace element analysis was employed to (1) assess the provenance of manufacture of Owasco vessels, and (2) determine whether compositional profiles correspond with assumptions about the reciprocal exchange of Owasco vessels in north-central Pennsylvania. Determining the provenance of manufacture required that the compositional profiles of Owasco and Clemson Island sherds as well as samples from five clay deposits be assessed. The results of this work suggest that the Owasco vessels found on these sites were locally manufactured and that Clemson Island and Owasco potters may have exploited similar deposits (Tables II and III). Of the 27 Clemson Island sherds analyzed, 18 grouped in the Tioga group, 4 in the West Branch group, and 5 in the Unknown group. Assuming that the Unknown group represents a clay deposit located in north-central Pennsylvania, all of the Clemson Island vessels appear to have been manufactured within north-central Pennsylvania.

Although all of the Clemson Island pots appear to have been manufactured in north-central Pennsylvania, there is evidence to suggest that pots were being exchanged between the region's occupants. Of the 27 Clemson Island sherds analyzed, 42% were manufactured from clays located more than 50 kilometers away. Evidence of this is reflected in the grouping of sherds from Fisher Farm and St. Anthony's in the Tioga group.

Analysis of the Owasco sherds indicates the use of similar clay deposits (Tables II and III). Of the 19 Owasco sherds analyzed, 11 grouped in the Tioga group, 4 in the Unknown group, and 4 in the Otsego group. Seventy-eight percent of the Owasco sherds appear to have been manufactured in north-central Pennsylvania. Four (21%) Owasco sherds exhibited compositional profiles similar to Clay samples 1 and 2 and probably represent non-locally manufactured containers. As I have argued elsewhere (*14*), given the limited number of non-locally manufactured pots found on these sites, it seems plausible that the pots themselves may not have been the objects of reciprocal exchange but rather their contents may have been important items that were moving between the populations of southern New York and north-central Pennsylvania. The pots that were recovered from these sites may represent disposed containers that were either broken in transit or were insufficient for holding their intended contents.

If the majority of the Owasco pots do not represent trade goods, how can we explain their presence at these sites? One possible explanation lies in the work of Hart (*7*) and others (*8*) who suggest that ceramic types represent artificial constructs whose main purpose is to perpetuate culture-history. The discontinuity between ceramic types and compositional profiles is difficult to explain within this framework since the use of types requires that the geographic and stylistic boundaries between groups be maintained reducing the likelihood that Owasco and Clemson Island populations could have exploited the same clay deposits.

This discontinuity can be viewed with less skepticism through population thinking. Employing a populationist framework allows stringent stylistic and geographic boundaries between groups to be removed allowing for the coexistance of different local populations within the same geographic area (*7*). Under this scheme, the limited differences between Clemson Island and Owasco populations have no reality but instead represent local differences within a single "Middle/Late Woodland population" (*24*). Evidence of the coexistence of different populations can be seen in the recovery of locally produced Owasco and Clemson Island sherds from the same features.

When viewed as a single population, the results of this project make sense and explain how Clemson Island and Owasco groups came to occupy the same ecological niche. When treated as a population, we should expect that potters would have made choices in the design and decoration of pots (*8*). While some choices may represent characteristics associated with the larger population, other choices may reflect the local population's interest in distinguishing itself from its neighbors. It is these individual choices and their variation between groups that make it difficult to apply a single culture-historic type to groups residing in the same region. Differences in the selection of one deposit over another probably reflect the range of materials that were available within a particular procurement zone.

Finally, the results of this project provide information about the relationship between raw material use and the manufacture of pots. Although many different deposits appear to have been used, there is some evidence to suggest that one or more high quality deposits may have been located and were repeatedly used by Native groups. Tight clustering of the data points in the West Branch group, suggest that a limited number of different clay deposits were exploited. These activities are consistent with what we know about the collection of resources among traditional potters. The clay's properties and its location are important criteria considered in the collection of raw materials (*9*). As demonstrated by Aaronson et al., although potters in their study procured materials from deposits located within a few kilometers of their settlement, potters indicated that they would change deposits if better clays could be located elsewhere. The repeated use of high quality deposits has also been demonstrated among other regional groups (*18, 25*).

Although one or more high quality deposits may have been located, it appears that several clay deposits within a smaller geographic region were exploited as represented by the loosely clustered data points in the Tioga group. The exploitation of numerous clay deposits within a restricted area is facilitated by the semi-permanent nature of these sites. Through the regular movement of camps and hamlets, potters may have encountered different resources at each location. There is also limited evidence to support a change in resource procurement between A.D. 900 and 1300. If such patterns were occurring, we would expect a change from the use of numerous clay deposits to a restricted number of deposits. Instead, as demonstrated by the Tioga Group, potters appear to have maintained earlier practices involving the exploitation of several different local deposits.

## Conclusion

Ceramic types belonging to the Early Late Prehistoric Clemson Island and Owasco types are often used to document and reconstruct interaction patterns in north-central Pennsylvania. Despite efforts to do so, the high degree of stylistic similarity between these ceramics often makes it difficult to distinguish between types limiting their use in interaction studies. In this study, trace element analysis was employed to determine if compositional profiles correspond with identified ceramic types. The study suggests that there is no clear distinction between ceramic types and exploited clay deposits. Instead, differences in the technological attributes of thes pots may reflect the location of sites to material resource zones.

## Acknowledgements

The artifacts used in this study were provided by Steven Warfel and Mark McConaghy of the Pennsylvania State Museum and Historical Commission in Harrisburg. William Lanford provided access to the x-ray fluorescence apparatus at the University at Albany, SUNY Accelerator Laboratoy. All errors are the sole responsibility of the author.

## References

1. Stewart, M. *Prehistoric Farmers of the Susquehanna Valley, Clemson Island Culture and the St. Anthony Site.* Occasional Publications in Northeastern Anthropology No. 13; Franklin Pierce College: Rindge, NH, 1994; pp 1-94.
2. Hatch, J. *The Fisher Farm Site, A Late Woodland Hamlet in Pennsylvania.* Occasional Papers in Anthropology, No. 12; The Pennsylvania University: University Park, PA, 1980; pp 1-74.
3. Snow, D. R. Migrations in Prehistory: The Northern Iroquoian Case. *American Antiquity* **1996**, 60, 59-80.
4. Schwartz, D. A Reevaluation of the Late Woodland Cultural Relationship in the Susquehanna Valley in Pennsylvania. *Man in the Northeast* **1985**, 29, 29-54.
5. Hay, C. A.; Hatch, J. W.; Sutton, J. A *Management Plan for Clemson Island Archaeological Resources in the Commonwealth of Pennsylvania.* Pennsylvania Historical and Museum Commission: Harrisburg, PA, 1987; pp 10-24.
6. Smith, I. In *Prehistoric Farmers of the Susquehanna Valley, Clemson Island Culture and the St. Anthony Site*; Stewart, R. M.; Occasional Publications in Northeastern Anthropology, No. 13; Archaeological Services: Bethlehem, CT, 1994; pp 13.
7. Hart, J. Another Look at Clemson's Island. *Northeast Anthropology* **1999**, 57, 19-26.

8. Chilton, E. In *The Archaeological Northeast*; Levine, M., Sassman, K. E., Nassaney, M.; Bergin and Garvey: Westport, CT, 1999; pp 97-111.
9. Aronson, M.; Skibo, J.M.; Stark, M.T.; In *Kalinga Ethnoarchaeology Expanding Archaeological Method and Theory*; Longacre, W.A.; Skibo, J.M.; Smithsonian Institution Press: Washington, DC, 1994; pp 83-112.
10. Arnold, D.E. *Ceramic Theory and Cultural Process*; Cambridge University Press: New York, NY, 1985; pp 20-94.
11. Lucy, C. Tioga Point Farm Sites 36BR3 and 36BR52: 1983 Excavations. *Pennsylvania Archaeologist* **1991**, 61, 1-19.
12. Lucy, C.; Vanderpoel, L.; The Tioga Point Site. *Pennsylvania Archaeologist* **1979**, 49, 1-12.
13. Lucy, C.; McCann, C.; The Wells Site, Assylum Township, Bradford County. *Pennsylvania Archaeologist* **1983**, 59, 1-12.
14. Rieth, C. In *Northeast Settlement and Subsistence Diversity, A.D. 700-1300*; Hart, J.P., Rieth, C.B.; New York State Museum Bulletin No. 496; University of the State of New York, Albany, New York, in press.
15. Neff, H.; Bove, F.J.; Lou, B; Piechowski, M.F. In *Chemical Characterization of Ceramic Pastes*; Neff, H.; Monographs in World Archaeology No. 7; Prehistory Press: Madison, WI, 1992; pp 93-112.
16. Trigger, B.G.; Yaffee, L.; Diksic, M.; Galinier, J.-L; Marshall, H; Pendergast, J.F. Trace-Element Analysis of Iroquoian Pottery. *Canadian Journal of Archaeology* **1980**, 4, 119-144.
17. Trigger, B.G.; Yaffee, L.; Dautet, D.; Marshall, H.; Pearce, R. Parker Festooned Pottery at the Lawson Site: Trace-Element Analysis. *Ontario Archaeology* **1984**, 42, 3-12.
18. Stimmell, C.; Pilon, J.; Hancock, R. In *Tools for Tomorrow, Archaeological Methods in the 21st Century, Proceedings of the 1991 Symposium*; Ottawa Chapter of the Ontario Archaeological Society: Ottawa, Ontario, Canada, 1991; pp 47-58.
19. Elam, J. M.; Carr, C.; Glascock, M.D.; Neff, H. In *Chemical Characterization of Ceramic Pastes*; Neff, H.; Monographs in World Archaeology No. 7; Prehistory Press: Madison, WI, 1992; pp 93-112.
20. Kuhn, R.D.; Sempowski, M.L. A New Approach to Dating the League of the Iroquois. *American Antiquity* **2001**, 66, 301-314.
21. Kuhn, R.D. Interaction Patterns in Eastern New York: A Trace Element Analysis of Iroquoian and Algonkian Ceramics. *The Bulletin and Journal of the New York State Archaeological Associaton* **1986**, 92, 9-21.
22. Van Espen, P.; Janssens, K.; Swenters, I.; *Version 3.0 AXIL X-Ray Analysis Software*. Department of Chemistry, University of Antwerp, U.I.A.
23. Norusis, M.J.; *SPSS/PC+ Statistics 4.0 for the IBM PC/XT/AT and PS/2*. SPSS Inc.; Chicago, IL.
24. Graybill, J. In New Approaches to Other Pasts; Kinsey, W.F., Moeller, R.W.; Archaeological Services: Bethlehem, CT, 1989: pp 51-59.
25. Lizee, J.M.; Neff, H.; Glascock, M.D. Clay Acquisition and Vessel Distribution Patterns: Neutron Activation Analysis of Late Windsor and Shantok Tradition Ceramics from Southern New England. *American Antiquity* **1994**, 60, 515-630.

## Chapter 14

# Contribution of Stable Isotope Analysis to Understanding Dietary Variation among the Maya

**Robert H. Tykot**

**Laboratory for Archaeological Science, Department of Anthropology, 4202 East Fowler Avenue, SOC 107, University of South Florida, Tampa, FL 33620**

Stable carbon and nitrogen isotope ratios in skeletal tissues are widely used as indicators of prehistoric human diets. This technique distinguishes between the consumption of C3 and C4 plants and assesses the contribution of aquatic resources to otherwise terrestrial diets. Isotopic ratios in bone collagen emphasize dietary protein; those in bone apatite and tooth enamel reflect the whole diet. Bone collagen and apatite represent average diet over the last several years of life, while tooth enamel represents diet during the age of crown formation. The isotopic analysis of all three tissues in individuals at Maya sites in Belize, Guatemala, Honduras and Mexico reveals variation in the importance of maize, a C4 plant, based on age, sex, status, and local ecological factors, as well as dramatic changes in subsistence patterns from the Preclassic to Postclassic periods. These results enable a tentative synthesis of the dynamic relationship between subsistence and sociopolitical developments in ancient Mesoamerica.

Corn (maize) was the single most important New World crop at the time of European contact, and is widely considered to have been fundamentally important to the rise of indigenous civilizations including the Inca, Aztec, and Maya. Of the triumvirate of maize, beans and squash, there is evidence that at least squash (*Cucurbita* spp.) was domesticated by 10,000 years ago (*1-2*), although domesticated beans (*Phaseolus* spp.) appear a few thousand years ago at most (*3*), and maize was first domesticated somewhere in between. Other plants were also probably cultivated several thousand years prior to the introduction of ceramics and village settlements in several regions of Central and South America (*4-5*). It is the origins and subsequent importance of maize to these early civilizations that remains the focus of much archaeological and biological research. The focus of this paper will be on the Maya, the cultural group who occupied Belize, Guatemala, and the Yucatan peninsula of Mexico from about 1500 BC to 1500 AD. Using isotopic and other evidence of human diets, Maya subsistence practices will be examined in the context of developments and adaptations in other cultural and environmental areas of Mesoamerica, from its frontier in the highlands of northern Mexico to El Salvador and Honduras in the south.

Maize is widely thought to have been domesticated in the Rio Balsas region of western Mexico (*6*), although the chronology of its origins and dispersal are less well understood. In the Tehuacán Valley, where maize cobs seemingly were associated with Coxcatlán phase (7000-5500 BP) levels, direct AMS dating on a dozen cobs produced dates no older than 4700 ± 110 BP (*7-9*). Two cobs from Guilá Naquitz have now been dated to 5400 BP and are the oldest maize macrofossils currently known (*10*), while maize pollen has been found in 6000 BP levels at San Andrés near the Gulf Coast of Tabasco (*11*), Pope *et al.* 2001), and maize phytoliths by about 5500 BP at the Aguadulce rockshelter in Panama (*12-13*).

Even in well documented archaeological excavations, however, the context and dating of archaeobotanical specimens may be problematic (*14*), and the identification of small numbers and certain forms of maize phytoliths has also been argued by some to be unreliable (*15-16*). In any event, even after its initial domestication, there was most likely considerable temporal and spatial variability in the incorporation of maize into existing subsistence patterns, and it is extremely difficult to measure its importance relative to other food resources since there is no way to balance the uneven representation and preservation of botanical and faunal remains in the archaeological record. The consumption of maize can, however, be quantified through stable isotope analysis of human skeletal remains (*17-20*).

## Principles of Stable Isotope Analysis

Carbon and nitrogen isotope ratios in human bone (and other tissues, when they are preserved) may be used to reconstruct prehistoric diet because of differential fractionation, between certain plant groups, of atmospheric carbon dioxide during photosynthesis and of nitrogen during fixation or absorption. There are two stable isotopes each of carbon ($^{12}C$, $^{13}C$) and nitrogen ($^{14}N$, $^{15}N$), with $^{12}C$ and $^{14}N$ by far the most common in nature. Small differences in the ratios of these isotopes ($^{12}C/^{13}C$, $^{15}N/^{14}N$) can be measured by isotope ratio mass spectrometry using samples smaller than 1 milligram. High precision isotope measurements are reported using the delta notation ($\delta^{13}C$, $\delta^{15}N$) relative to the internationally recognized PDB and AIR standards and are expressed in parts per thousand or per mil (‰).

Empirical and experimental data have indicated that different bone tissues reflect different components of the diet (*21-22*). In general, bone collagen is produced primarily from the protein portion of the diet, while bone apatite and tooth enamel are produced from a mixture of dietary protein, carbohydrates and fats. Stable isotope analysis of both bone collagen and apatite thus permits quantitative estimates of several dietary components. Both bone collagen and apatite are constantly being resorbed and replenished, so that their isotopic composition reflects dietary averages over at least the last several years of an individual's life. The apatite in tooth enamel, however, is not turned over, and represents diet at the time of tooth formation, often from a pre-weaning age.

Typically, grasses originally native to hot, arid environments follow the C4 (Hatch-Slack) photosynthetic pathway, and will have $\delta^{13}C$ values averaging about -12‰; trees, shrubs, and grasses from temperate regions, which follow the C3 (Calvin-Benson) photosynthetic pathway, will have $\delta^{13}C$ values averaging about -26‰. Metabolic fractionation in their consumers results in bone collagen $\delta^{13}C$ values of about -7‰ and -21‰ and bone apatite $\delta^{13}C$ values of about 0‰ and -14‰, respectively, for pure C4 and pure C3 diets.

Stable carbon isotope analysis is particularly useful in New World dietary studies since maize is often the only C4 plant contributing significantly to human diets; its contribution to bone collagen and to bone apatite may be estimated by interpolation. Some caution is warranted, however, if succulent plants were present, since they utilize the alternative CAM (crassulacean acid metabolism) photosynthetic pathway which results in carbon isotope ratios similar to those of C4 plants. Nevertheless, CAM plants are unlikely to have been major sources of dietary protein, whether consumed directly or indirectly through herbivorous faunal intermediaries.

While most plants C3 and C4 plants have similar carbon isotope ratios in most ecological settings, care must be taken in forested or other environments where atmospheric recycling can occur resulting in depleted carbon isotope ratios in plants and their consumers (*23*). Human diets which include forest animals, therefore, can significantly mask the consumption of maize or other C4 plants with enriched carbon isotope ratios if the values for these food resources have not been established. Carbon isotope ratios in C3 plants also are affected by water stress and thus correlate to altitude and other factors. This is an important factor to consider in establishing the isotopic baseline for C3 plant foods; C4 plants, with their photosynthetic adaptation to hot and arid environments, do not seem to vary to the same extent, so that a carbon isotope ratio of about -10‰ for maize may be used throughout Mesoamerica unless demonstrated otherwise.

The carbon isotope ratios of marine and freshwater organisms are more variable, depending on local ecological circumstances, and often overlap with those of terrestrial plants and their consumers. These foods typically have much higher nitrogen isotope values, however, and their high protein content will contribute much more carbon to bone collagen than will plants foods which are only 10-20% protein.

For all of these reasons, quantification of the contribution of maize or other plants to human diets is better accomplished through measurement of stable isotope ratios in bone apatite or tooth enamel, rather than bone collagen.

The nitrogen isotope ratios for plants depend primarily on how they obtain their nitrogen - by symbiotic bacterial fixation of atmospheric nitrogen, or from soil nitrates - and these values are similarly passed along through the food chain accompanied by an approximately 2-3‰ positive shift for each trophic level. Human consumers of terrestrial plants and animals typically have $\delta^{15}N$ values in bone collagen of about 6-10‰ whereas dedicated consumers of freshwater or marine fish, and sea mammals, may have $\delta^{15}N$ values of 15-20‰ (*24*). Nitrogen isotope ratios also vary according to rainfall and other factors (*25*), and both carbon and nitrogen isotope ratios vary considerably among marine organisms (*26*).

It is critical, therefore, to establish a site-relevant isotopic baseline for interpreting human skeletal data. Analyses of faunal remains provide a good estimate both of the animals themselves and the plants they consume. This is particularly important since in many areas while maize may be the only C4 plant cultivated and consumed directly by humans, other C4 grasses may exist and be consumed by grazing herbivores which pass on an enriched isotopic signature to their human consumers. Information from ethnohistoric and other sources, along with isotopic data for relevant food resources, may then be used to propose specific dietary mixing models to account for human bone isotope ratios (*27*).

## Analytical Procedures

For bone chemistry studies in Mesoamerica, the extraction of collagen and apatite from bone is often problematic due to rapid degradation in the warm and moist climate. Procedures must be employed which remove carbon and nitrogen from non-biogenic sources, and assess whether the remaining material is sufficiently intact or may have fractionated during decomposition. Bone collagen is usually obtained by demineralization with HCl or EDTA, followed by treatment with NaOH to remove humic acid contaminants, and with a methanol, chloroform, and water mixture to remove lipids. The use of weak HCl (2%) frequently allows the recovery of collagen from otherwise highly degraded bone, and the visible collagen pseudomorphs produced are a strong indicator of sample reliability. The resulting collagen is often then solubilized, particulate organic contaminants removed by filtration or centrifugation, and recovered as a gelatin powder following evaporation (*27-28*).

$CO_2$ and $N_2$ gases are subsequently obtained and introduced to a stable isotope ratio mass spectrometer either (1) by off-line combustion in evacuated quartz tubes, followed by purification and separation by cryogenic distillation, and the use of a manifold 'cracker'; or (2) by on-line combustion and temporal separation of the gases using a CHN analyzer and chromatographic column directly coupled to the mass spectrometer. The reliability of collagen samples is evaluated by the yield of 'collagen' from the bone sample, and the ratio of C:N in the 'collagen'. Yields under 1% signify very poor preservation and the likelihood that the remaining material may have become fractionated, while ratios outside the range of 2.9-3.7 are not considered typical for intact collagen (*29*).

Although normal organic decomposition is the greatest problem affecting bone collagen, contamination and chemical exchange with the burial matrix are the main issues for the reliable isotopic analysis of the carbonate in bone apatite and tooth enamel (*30-32*). The denser, relatively impermeable tooth enamel, however, is easily dealt with even when the tooth is of great antiquity by removal of surface layers followed by an acetic acid wash (*33-34*). Similar procedures have been developed for the more porous bone apatite, with residual organic components dissolved in sodium hypochlorite and non-biogenic carbonates removed with a buffered acetic acid solution (*35*). Samples may be further tested for integrity by x-ray diffraction or IR spectroscopy. Purified $CO_2$ is obtained by reaction with 100% phosphoric acid, either off-line (followed by cryogenic distillation and introduction through a manifold system as described above), or online in automated acid bath systems.

High precision in stable isotope mass spectrometry is achieved through repeated measurements of a reference gas, and inter-laboratory compatibility is maintained through calibration against standard reference materials. Analytical precision for most modern mass spectrometers is about ± 0.1‰ for both $\delta^{13}C$ and $\delta^{15}N$.

## Maya Isotope Studies

One of the first applications of stable isotope analysis in archaeology was the analysis of human remains from the Tehuacan Valley (*36*). Unfortunately, of the 12 samples which produced collagen, only two came from contexts earlier than 3000 BP. These limited samples, however, suggest that maize was already a dietary staple several millennia earlier in the El Riego phase ($\delta^{13}C$ = -13.3‰), and dominated the diet by the succeeding Coxcatlan phase ($\delta^{13}C$ = -6.1‰) (*37*). The widespread importance of maize by the Formative Period at villages like Monte Alban in the Oaxaca Valley, and along the southeastern coast of Mexico in the Soconusco region, has also been documented by stable isotope studies (*38-39*). At the southern end of Mesoamerica, maize appears to have been a significant component of Panamanian diets by 4500 BP, at preceramic coastal sites such as Cerro Mangote (*40*).

A significant amount of research on the diet of the Maya has been done in the last dozen years since the first published study in this central region of Mesoamerica (*41*). Results are now available for over 600 individuals, representing such diverse areas as the lowlands of Belize and the Petén, the Yucatan peninsula, and the highlands of Guatemala, and spanning more than 2000 years from the Preclassic through Postclassic periods (Figure 1; Table 1). It is now well established that substantial variation in subsistence adaptations existed in this region, and this diversity is mainly attributable to chronology, geographic location, and social factors (*42-45*). The importance of maize generally increases over time, probably due to intensification and extensification of agricultural production; communities living in diverse ecological zones nevertheless employed different subsistence strategies; and the increasingly hierarchical and complex nature of Maya society allowed for differential patterns of food acquisition, storage, and consumption.

In the Maya region, zooarchaeological and paleobotanical remains as well as ethnohistoric evidence have identified the most important animal and plant foods as including white-tailed and other deer, domestic dog, armadillo, peccary, and freshwater turtle, as well as rabbit, opossum, raccoon, agouti, and tapir; fish and shellfish (in coastal and estuary areas); and maize, beans, squash, root crops, and fruits (*46-47*). CAM plants such as cactus and pineapple are not thought to have been important to the diet. Trace element analyses strongly argue that most foods were obtained locally, rather than by trade (*48*), although by the Classic period elites likely would have had preferential access to non-local resources (*49*).

Most notably, dog was the only domesticated animal available for consumption, and hunting of wild animals remained a significant part of Maya subsistence adaptations long after the establishment of settled villages and urban development. Stable isotope analyses of dog remains from a number of Maya sites generally show similar values to their human masters, but some, especially in later periods, have even higher maize signatures indicating that they were specifically being fed large quantities of maize (*42, 45, 50-52*). Finds of dog remains in domestic middens, often with cutmarks on the bones, demonstrates that these were indeed comestible canids or corn-dogs (*53*). Isotopic analyses of deer collagen, however, have revealed at most a 7-22% level of maize consumption, attributable to occasional browsing in maize fields, and this level of maize consumption does not change over time (*42, 45, 52, 54-55*).

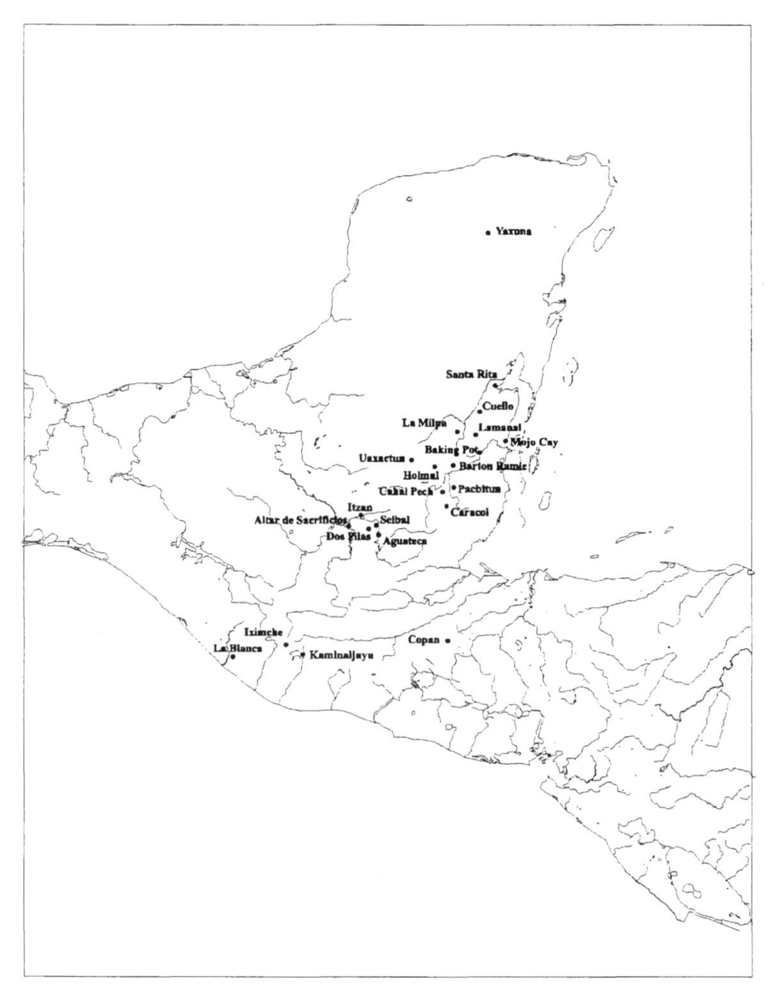


*Figure 1. Map showing Maya archaeological sites for which stable isotope data are available.*

**Table I. Stable Isotope Data for the Maya**

| Site | n | $\delta^{15}N$ | $\delta^{13}C_{collagen}$ | $\delta^{13}C_{apatite}$ | $\delta_{13}C_{enamel}$ | Ref |
|---|---|---|---|---|---|---|
| *Belize* | | | | | | |
| Baking Pot: LC | 9,9,4 | 9.2±1.3 | -11.0±1.0 | -6.6±0.6 | | *50* |
| Barton Ramie: EC | 7,7,6 | 8.6±0.5 | -11.4±0.9 | -7.4±0.6 | | *50* |
| Barton Ramie: LC | 31,31,24 | 8.9±0.4 | -11.2±1.5 | -6.9±0.5 | | *50* |
| Cahal Pech: PreC | 11,11 | 9.4±1.1 | -11.7±1.5 | | | *58* |
| Caracol: EC | 8,8 | 9.0 | -9.1 | | | *59-61* |
| Caracol: LC | 72,72 | 9.5 | -10.0 | | | *59-61* |
| Caracol: TC | 5,5 | 9.6 | -10.8 | | | *59-61* |
| Cuello: PreC | 23,28,16,33 | 8.9±1.0 | -12.9±0.9 | -9.8±1.0 | -8.7±2.3 | *42,45* |
| Lamanai: PreC | 2,2,3,1 | 10.2 | -12.7 | -6.9±0.2 | -8.8 | *41,70* |
| Lamanai: EC | 4,4,4 | 10.9±1.3 | -12.3±1.4 | -6.4±1.4 | | *41,70* |
| Lamanai: LC | 2,3,4 | 10.4 | -14.1±0.9 | -6.1±0.8 | | *41,70* |
| Lamanai: TC | 7,7,4,1 | 9.9±0.4 | -15.0±1.1 | -7.4±1.2 | -7.3 | *41,70* |
| Lamanai: PC | 24,24,18,2 | 9.5±0.9 | -9.3±0.8 | -6.4±1.7 | -2.0 | *41,70* |
| Lamanai: H | 10,11,9,1 | 9.7±0.6 | -9.5±0.6 | -5.6±0.7 | -1.8 | *41,70* |
| Mojo Cay: MC | 8,8 | 10.1±0.9 | -8.5±0.4 | | | *56* |
| Pacbitun: EC | 1,1,1 | 8.1 | -9.2 | -5.1 | | *62,70* |
| Pacbitun: LC | 3,3,3 | 9.3±0.5 | -8.5±1.1 | -4.7±0.6 | | *62,70* |
| Pacbitun: TC | 16,16,14,3 | 9.3±0.7 | -10.6±1.4 | -5.9±1.0 | -5.6±1.4 | *62,70* |
| *Peten* | | | | | | |
| Aguateca: LC | 7,8 | 9.4±1.0 | -9.6±0.6 | | | *51* |
| Altar: PreC | 9,9 | 8.2±0.9 | -10.7±1.1 | | | *51* |
| Altar: EC | 10,10 | 8.4±0.4 | -9.2±0.4 | | | *50-51* |
| Altar: LC | 18,19 | 9.1±0.9 | -8.8±0.8 | | | *50-51* |
| Altar: TC | 16,16 | 8.8±1.1 | -9.0±0.9 | | | *51* |
| Dos Pilas: LC | 15,15 | 9.8±0.9 | -9.0±1.0 | | | *51* |
| Dos Pilas: TC | 4,4 | 8.8±1.0 | -9.4±0.7 | | | *51* |
| Holmul: EC | 13,12,18 | 9.4±0.9 | -9.5±1.3 | -4.5±1.3 | | *50* |
| Holmul: LC | 2,2,4 | 9.1 | -9.0 | -4.0±0.7 | | *50* |
| Itzan: LC | 5,5 | 8.0±0.9 | -9.2±0.3 | | | *51* |
| Seibal: PreC | 7,7 | 9.7±0.7 | -9.6±0.9 | | | *51* |
| Seibal: EC | 3,3,3 | 10.0±0.9 | -10.2±0.5 | -6.4±0.6 | | *50* |
| Seibal: LC | 37,24,24 | 9.6±0.7 | -9.6±1.3 | -6.2±0.9 | | *50-51* |
| Seibal: TC | 13,13 | 8.6±0.7 | -9.4±1.1 | | | *51* |
| Uaxactun: LC | 5,6,2 | 9.4±0.9 | -10.7±1.0 | -5.7 | | *50* |
| *Honduras* | | | | | | |
| Copan: EC | 2,2,2 | 6.8 | -11.0 | -4.9 | | *50* |
| Copan: LC | 37,39,34 | 7.6±0.8 | -10.2±0.9 | -5.6±1.0 | | *50* |
| Copan: Classic | 46,46 | 7.6±0.5 | -9.3±0.7 | | | *64* |
| *Guatemala* | | | | | | |
| Iximche: PC | 13,13,43 | 7.9±0.4 | -7.8±0.4 | | -2.1±1.1 | *65,71* |
| Kaminaljuyu: EC | 15,24,96 | 9.2±1.8 | -9.9±1.0 | | -3.0±1.4 | *67-68* |
| La Blanca: PreC | 3,3 | 9.1±0.3 | -12.5±1.2 | | | *38* |

Abbreviations: Preclassic (PreC); Early Classic (EC); Middle Classic (MC); Late Classic (LC); Terminal Classic (TC); Postclassic (PC); Historic (H)

For the Maya lowlands in Belize, collagen data are now available for ten sites, including one located on the coast (*41-42, 45, 50, 56-62*) (Figure 2). These data show that by the Preclassic period, with $\delta^{13}C$ values at non-coastal sites averaging -12.6 ± 1.2‰, C4 sources appear to account for about 50% of the carbon in bone collagen; this is likely the result of consuming significant quantities of maize, as well as C4-enriched animal foods including dog and armadillo. Surprisingly, there is only a modest increase in isotope ratios throughout the Early, Late, and Terminal Classic periods, to -11.3 ± 2.3‰, suggesting that the diversity of foods available in the Maya lowlands had not significantly changed over 2000 years.

It must be noted that the figures presented here are averages of data from multiple sites with different local ecological settings; not all sites individually follow the same trend, for example at Caracol where carbon isotope ratios decrease rather than increase over the course of the Classic period (*59-61*). The Baking Pot, Barton Ramie and Lamanai sites are located in riverine settings, where consumption of freshwater fish with depleted carbon isotope ratios could offset C4 maize contributions to bone collagen. At Mojo Cay, the enriched carbon isotope ratios are most likely due not to high levels of maize consumption, but rather to the importance of local reef fish and shellfish which do not have enriched nitrogen isotope ratios (*42, 45, 56*). Intra-site differences based on status are apparent at sites such as La Milpa (unpublished data), Pacbitun (*62*) and Caracol (*59-61*), with elites buried in formal tombs near site centers enriched 2-4‰ in collagen carbon compared to non-elites buried away from the site core. At Lamanai, Early Classic elites have the lowest carbon isotope ratios, indicating the least dependence on maize and probably preferential access to reef fish and shellfish (*41*). Of great significance is the observed shift at Lamanai between the Classic and Postclassic periods, to $\delta^{13}C$ values of -9.3 ± 0.8‰, demonstrating a substantially increased reliance on maize and possibly also a shift from the consumption of freshwater to marine fish as part of coastal trade networks (*41*).

At Preclassic Maya sites in the Peten region like Altar de Sacrificios and Seibal the average collagen $\delta^{13}C$ values of -10.2 ± 1.2‰ suggest that C4 sources accounted for about 70% of the carbon in bone collagen, a level only reached in Belize during the Postclassic (Figure 3). During the course of the ensuing Classic period, however, we observe the same modest increase of about 1‰ in $\delta^{13}C$ values for Peten sites as we saw for Belize; intensification of maize production was something which was occurring throughout the Maya region. There is no observable difference in carbon isotope ratios between Peten sites located in riverine and inland environments, suggesting that fish were not a significant component of the average Maya diet in this region. At Seibal, there is no change in collagen carbon isotope ratios from the Preclassic to the Terminal classic, although nitrogen isotope ratios decrease nearly 1‰, suggesting stability in maize consumption at this site while terrestrial fauna may have become more important than aquatic fauna over time

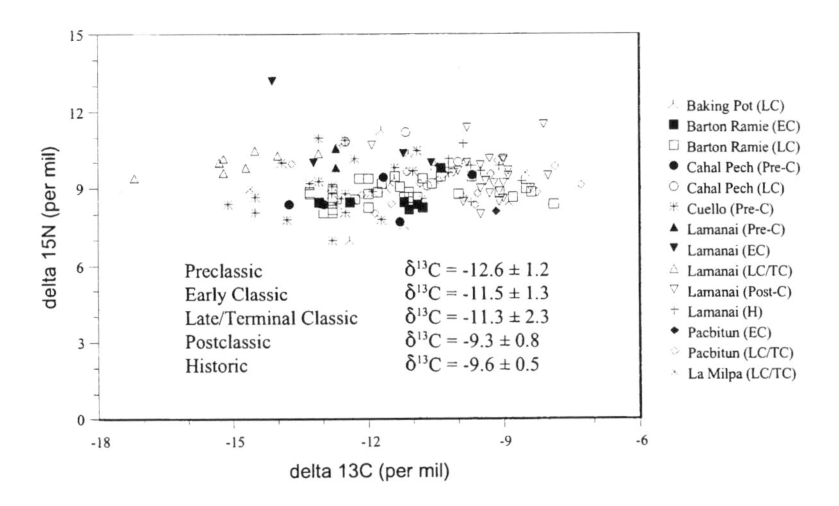


*Figure 2. Bone collagen stable isotope data for Maya sites in Belize. Mojo Cay (coastal) not shown.*

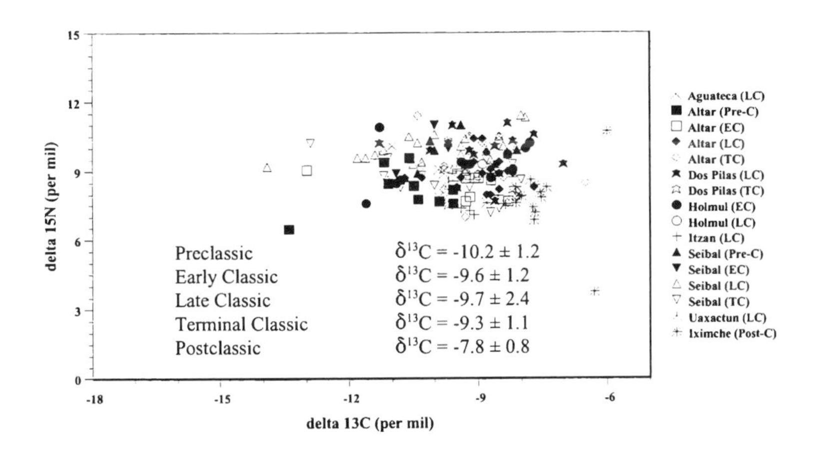


*Figure 3. Bone collagen stable isotope data for Maya sites in the Peten and Guatemala.*

(*51*). Individuals from Late Classic Uaxactun notably more closely resemble contemporaries in Belize in their isotope ratios, perhaps because all of the sampled individuals are females from an elite burial context (*50*). At Copan in Honduras, carbon isotope ratios are similar to contemporary Classic sites in the Peten region; nitrogen isotope ratios, however, are considerably lower, suggesting then greater dependence on beans and/or reduced availability of animal protein (*63*). Greater variability in isotope ratios is also observed among high-status burial groups, perhaps a result of greater diversity in diets (*64*). While nitrogen isotope ratios could also be lower at Copan because of its higher altitude (ca. 600 m), there is no systematic difference in the $\delta^{15}N$ values for white-tailed deer and other fauna from Copan relative to Peten sites (*50*). The only Postclassic site for which data are available is Iximche, the Kaqchikel Maya capital in highland Guatemala, with $\delta^{13}C$ values reaching -7.8 ± 0.8‰ (*65-66*), indicating substantially greater maize dependence than at any Classic period site, or contemporary Lamanai in Belize. No faunal samples from this highland area (2200 m asl) have been tested to determine whether the accompanying low nitrogen isotope ratios (7.9 ± 0.4‰) are a result of environmental factors or greater dependence on beans or other plant foods. The collagen data listed in Table 1 for Kaminaljuyu actually come from tooth dentin, with some of the teeth having enriched carbon isotope ratios as a result of the pre-weaning diets that they represent (*67-68*). This only further emphasizes the differences between Classic and Postclassic diets in the Peten and highland Guatemala.

For omnivorous humans, it is important to remember that isotope data from bone collagen are heavily biased towards consumed protein, making it difficult to compare absolute quantities of maize consumed when the amount of animal (or fish) protein also varied. Bone apatite and tooth enamel, on the other hand, appear to faithfully reflect the whole diet (*21*), allowing the calculation of the carbon isotope ratios of the average foods consumed. Along with nitrogen isotope ratios from bone collagen, estimated human diets may then be illustrated in the same plot as the foods themselves. Carbon isotope ratios in bone apatite and tooth enamel ratios are offset here by -9.5‰ based on the controlled diet experiments on rats by Ambrose and Norr (*21*), although an offset of -12‰ has been empirically determined for larger mammals (*69*); nitrogen isotope ratios are offset by -3.0‰. Values for animal flesh, which contains both protein and fats, were estimated by a -2.0‰ correction relative to the bone collagen results for the faunal specimens tested; when modern samples were used, these were corrected +1.5‰ to account for changes in atmospheric $CO_2$ due to combustion of fossil fuels since the industrial revolution (Figures 4-5).

Bone apatite data are available for a half dozen sites in Belize, and show a similar chronological trend towards increasingly enriched carbon isotope ratios as that observed for bone collagen. A large shift in $\delta^{13}C$ from -9.5 ± 1.2‰ to -6.8 ± 1.2‰, however, is observed between the Preclassic and Early Classic at sites

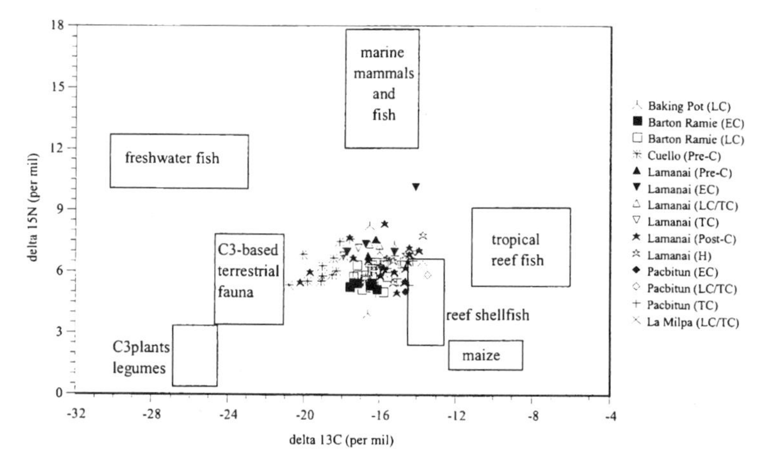


*Figure 4. Human diets at Maya sites in Belize based on bone apatite carbon and bone collagen nitrogen isotope ratios. Human bone apatite carbon corrected -9.5, and collagen nitrogen -3.0 to reflect estimated diet. Faunal bone collagen carbon corrected -2.0 to simulate flesh; modern samples corrected +1.5 for industrial effect.*

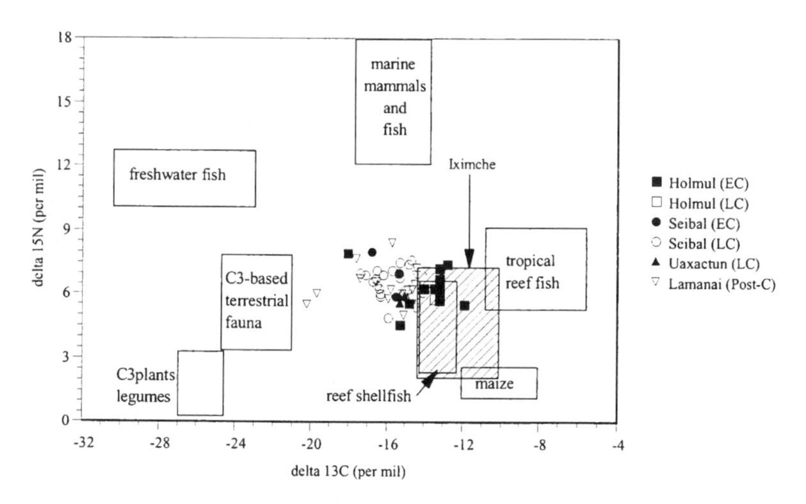


*Figure 5. Human diets at Maya sites in the Peten and Guatemala based on bone apatite carbon and bone collagen nitrogen isotope ratios. Corrections as in Figure 4.*

including Cuello and Lamanai, while only a statistically insignificant change is then observed through the Late, Terminal, and Postclassic periods. This is in contrast to the timing of changes in carbon isotope ratios in bone collagen, and suggests that the real increase in the importance of maize to the whole diet came during the first millennium BC, and that the collagen shift between the Classic and Postclassic periods may be attributed to changes in the average carbon isotope value for dietary protein. This could be explained by a decrease in the consumption of animal foods, thereby increasing the relative contribution of maize protein; eating animals with more enriched carbon isotope ratios such as dogs, or reef fish if available; or some combination of these.

For the Classic period sites of Holmul, Seibal, and Uaxactun in the Peten region, and Copan in Honduras, there do not appear to be significant differences in the carbon isotope ratios of bone apatite relative to the BelizeClassic period sites, with $\delta^{13}C$ values ranging from about -4 to -6‰, suggesting that the overall importance of maize was actually quite similar throughout these regions. Tooth enamel samples from Early Classic Kaminaljuyu average $\delta^{13}C$ of -3.0 ±1.4‰, but the teeth tested include molars and premolars formed at both pre- and post-weaning ages, so that this average value is enriched due to the trophic level effect; the almost-adult diets represented by 3$^{rd}$ molars, with their enamel formed between ages 10 and 12, are therefore similar in isotopic value to those observed for Belize and the Peten. A second and a third molar from Lamanai, which average -2.0‰, are not consistent with bone apatite values from the same individuals and may be the result of small samples representing short-term seasonal diets at the time of tooth formation (*70*). For Postclassic Iximche, however, only third molars were tested, and the $\delta^{13}C$ values of -2.1 ± 1.1‰ obtained are clearly enriched relative to Classic Maya sites in all regions (*71*). The only contemporary site available for comparison is Lamanai, where bone apatite $\delta^{13}C$ averages -6.4±1.7‰, thus strongly supporting the conclusion that dependence on maize was significantly greater in the Mesoamerican highlands than in the lowlands.

## Conclusion

Ecological models of the 'collapse' of Maya civilization have argued that dietary and health deficiencies mounted throughout the Classic period due to increasing urbanization, deforestation, and climatic instability (*72-73*), but no pattern of nutritional or health deterioration over time has been documented (*43, 74-78*). Nevertheless, increasing demands of a political elite certainly encouraged agricultural intensification, monocropping, and unsustainable extensification and use of marginal lands and this is clearly evident in changes in bone isotope ratios. Notable shifts occur in bone apatite values between the Preclassic and Early Classic periods, and in bone collagen between the Late/Terminal Classic and Postclassic periods, suggesting that overall dependence on maize was largely established and stable by the beginning of the Classic, and that changes in protein sources occurred after the Classic Maya collapse. Decreasing availability of limited wild animal populations may be hypothesized as contributing to increased territorial conflicts among Maya centers. Furthermore, significant variation is apparent on regional levels, with inland and highland populations more dependent on maize in all periods relative to lowland sites with access to aquatic resources or located in more diverse ecological settings. Elite privilege was also important in determining diets during the Classic period, but maize or maize-fed animals were not always the preferred food despite their important role in Maya rituals.

The synthesis presented here is subject to revision as new data become available, both isotopic and from other sources. Results are currently available from only two post-Classic sites, and sites in Honduras, highland Guatemala, and the Yucatan are very sparsely represented. Almost no data are available from the adjacent gulf coast region of Mexico, where the Olmec civilization developed.

Methodologically, bone apatite needs to be analyzed for many more sites to avoid the protein bias of bone collagen, when we know that plant foods formed the bulk of non-coastal Mesoamerican diets. Measurement of isotope ratios in teeth of various formation ages has already been used to reconstruct weaning practices, and in comparison to adult values, changes in diet due to migration or other factors. On a finer scale, microsampling of growth layers in tooth enamel and dentin can also be used to document seasonal variability in diets, which previously was limited to studies in areas where hair samples are preserved. Finally, compound specific isotopic analysis is a new and promising technique which allows comparison of essential vs. non-essential amino acids in collagen, while eliminating the problem of diagenesis in poorly preserved bones.

The continued integration of bioarchaeological data with ethnohistoric, archaeological, faunal, floral, and paleoclimatic evidence ultimately will allow more precise interpretations of the relationships between subsistence and sociopolitical developments in ancient Mesoamerica.

## References


1. Smith, B. D. *Science* **1997**, *276(5314)*, 932-934.
2. Piperno, D. R.; Andres, T. C.; Stothert, K. E. *Journal of Archaeological Science* **2000**, *27*, 193-208.
3. Kaplan, L.; Lynch, T. *Economic Botany* **1999**, *53*, 261-272.
4. Piperno, D. R.; Holst, I. *Journal of Archaeological Science* **1998**, *25*, 765-76.
5. Piperno, D. R.; Pearsall, D. M. *The Origins of Agriculture in the Lowland Neotropics*. Academic Press: London, 1998.
6. MacNeish, R. S.; Eubanks, M. W. *Latin American Antiquity* **2000**, *11*, 3-20.
7. Long, A.; Benz, B.; Donahue, J.; Jull, A.; Toolin, L. *Radiocarbon* **1989**, *31*, 1035-40.
8. MacNeish, R. S. *Latin American Antiquity* **2001**, *12*, 99-104.
9. Long, A.; Fritz, G. J. *Latin American Antiquity* **2001**, *12*, 87-90.

10. Piperno, D. R.; Flannery, K. V. *Proceedings of the National Academy of Sciences* **2001**, *98*, 2101-2103.
11. Pope, K. O.; Pohl, M.; Jones, J. G.; Lentz, D. L.; von Nagy, C.; Vega, F. J.; Quitmyer, I. R. *Science* **2001**, *292*, 1370-1373.
12. Piperno, D. R.; Clary, K. H.; Cooke, R. G.; Ranere, A. J.; Weiland, D. *American Anthropologist* **1985**, *87*, 871-78.
13. Piperno, D. R.; Holst, I.; Ranere, A. J.; Hansell, P.; Stothert, K. E. *The Phytolitharien* **2001**, *13 (2&3)*, 1-7.
14. Rossen, J.; Dillehay, T. D.; Ugent, D. *Journal of Archaeological Science* **1996**, *23*, 391-407.
15. Rovner, I. Phytolith analysis. *Science* **1999**, *283(5406)*, 488-489.
16. Russ, J. C. Rovner, I. *American Antiquity* **1989**, *54*, 784-792.
17. van der Merwe, N. J.; Vogel, J. C. *American Antiquity* **1977**, *42*, 233-242.
18. van der Merwe, N. J. *Proceedings of the British Academy* **1992**, *77*, 247-264.
19. Schwarcz, H. P.; Schoeninger, M. J. *Yearbook of Physical Anthropology* **1991**, *34*, 283-321.
20. Katzenberg, M. A. In *Biological Anthropology of the Human Skeleton*; Katzenberg, M. A.; Saunders, S. R., Eds.; Wiley-Liss: New York, 2000; pp 305-327.
21. Ambrose, S. H.; Norr, L. In *Prehistoric Human Bone. Archaeology at the Molecular Level*; Lambert, J. B.; Grupe, G., Eds.; Springer Verlag: Berlin, 1993; pp 1-37.
22. Tieszen, L. L.; Fagre, T. In *Prehistoric Human Bone: Archaeology at the Molecular Level*; Lambert, J. B.; Grupe, G., Eds.; Springer-Verlag: New York, 1993; pp 121-155.
23. van der Merwe, N. J.; Medina, E. *Journal of Archaeological Science* **1991**, *18*, 249-59.
24. Schoeninger, M. J.; DeNiro, M. J. *Science* **1983**, *220*, 1381-1383.
25. Ambrose, S. H. *Journal of Archaeological Science* **1991**, *18*, 293-317.
26. Schoeninger, M. J.; DeNiro, M. J. *Geochimica et Cosmochimica Acta* **1984**, *48*, 625-639.
27. Little, J. D. C.; Little, E. A. *Journal of Archaeological Science* **1997**, *24*, 741-747.
27. Ambrose, S. H. *Journal of Archaeological Science* **1990**, *17*, 431-51.
28. Ambrose, S. H. In *Investigations of Ancient Human Tissue: Chemical Analyses in Anthropology*; Sandford, M. K. Ed.; Gordon and Breach Science Publishers: Langhorne, PA, 1993; pp 59-130.
29. DeNiro, M. J. *Nature* **1985**, *317*, 806-809.
30. Schoeninger, M. J.; DeNiro, M. J. *Nature* **1982**, *297*, 577-578.
31. Sullivan, C. H.; Krueger, H. W. *Nature* **1983**, *301*, 177.
32. Krueger, H. W. *Journal of Archaeological Science* **1991**, *18*, 355-361.
33. Lee-Thorp, J. A.; van der Merwe, N. J. *South African Journal of Science* **1987**, *83*, 712-15.
34. Lee-Thorp, J. A. In *Biogeochemical Approaches to Paleodietary Analysis*; Ambrose, S. H.; Katzenberg, M. A., Eds.; Plenum: New York, 2000; pp 89-116.
35. Koch, P. L.; Tuross, N.; Fogel, M. L. *Journal of Archaeological Science* **1997**, *24*, 417-429.

36. DeNiro, M. J.; Epstein, S. *Geochimica et Cosmochimica Acta* **1981**, *45*, 341-351.
37. Farnsworth, P.; Brady, J. E.; DeNiro, M.; MacNeish, R. S. *American Antiquity* **1985**, *50*, 102-116.
38. Blake, M., Chisholm, B. S., Clark, J. E., Voorhies, B. & Love, M. W. *Current Anthropology* **1992**, *33*, 83-94.
39. Blitz, J. PhD dissertation, University of Wisconsin-Madison, WI, 1995.
40. Norr, L. In *Archaeology in the Lowland American Tropics*; Stahl, P. W., Ed.; Cambridge University Press: Cambridge, England, 1995; pp 198-223.
41. White, C. D.; Schwarcz, H. P. *Journal of Archaeological Science* **1989**, *16*, 451-74.
42. Tykot, R. H.; van der Merwe, N. J.; Hammond, N. In *Archaeological Chemistry: Organic, Inorganic and Biochemical Analysis;* Orna, M. V., Ed.; American Chemical Society: Washington, DC, 1996; pp 355-365.
43. Wright, L.E.; White, C.D. *Journal of World Prehistory* **1996**, *10*, 147-198.
44. Gerry, J. P.; Krueger, H. W. In *Bones of the Maya: Studies of Ancient Skeletons*; Whittington, S. L.; Reed, D. M., Eds.; Smithsonian Institution Press: Washington, DC, 1997; pp 196-207.
45. van der Merwe, N. J.; Tykot, R. H.; Hammond, N.; Oakberg, K. In *Biogeochemical Approaches to Paleodietary Analysis*; Ambrose, S. H.; Katzenberg, M. A., Eds.; Plenum: New York, 2000; pp 23-38.
46. Reed, D. M. In *Reconstructing Ancient Maya Diet*; White, C.D., Ed.; University of Utah Press: Salt Lake City, 1999; pp 183-196.
47. Lentz, D. L. In *Reconstructing Ancient Maya Diet*; White, C.D., Ed.; University of Utah Press: Salt Lake City, 1999; pp 3-18.
48. Wright, L. E. In *Reconstructing Ancient Maya Diet*; White, C.D., Ed.; University of Utah Press: Salt Lake City, 1999; pp 197-220.
49. Gerry, J. P. *Geoarchaeology* **1997**, *12*, 41-69.
50. Gerry, J. P. PhD Dissertation, Harvard University, Cambridge, MA, 1993.
51. Wright, L. E. PhD dissertation, University of Chicago, Chicago, IL, 1994.
52. White, C. D.; Pohl, M. E.; Schwarcz, H. P.; Longstaffe, F. *Journal of Archaeological Science* **2001**, *28*, 89-107.
53. Clutton-Brock, J.; Hammond, N. *Journal of Archaeological Science* **1994**, *21*, 819-826.
54. Emery, K. F. In *Reconstructing Ancient Maya Diet*; White, C.D., Ed.; University of Utah Press: Salt Lake City, 1999; pp 83-102.
55. Emery, K.; et al. *Journal of Archaeological Science* **2000**, *27*.
56. Norr, L. C. PhD dissertation, University of Illinois, Urbana-Champaign, IL, 1991.
57. Mansell, E.; Tykot, R. H.; Hammond, N. Paper presented at the Society for American Archaeology Annual Meeting, 2000.
58. Powis, T. G.; Stanchly, N.; White, C. D.; Healy, P. F.; Awe, J. J.; Longstaffe, F. *Antiquity* **1999**, *73*, 364-376.
59. Chase, A. F.; Chase, D. Z. Paper presented at the 14th International Congress of Anthropological and Ethnological Sciences, Williamsburg, VA, 1998.
60. Chase, A. F.; Chase, D. Z. In *Royal Courts of the Ancient Maya. Volume 2: Data and Case Studies*; Inomata, T.; Houston, S. D., Eds.; Westview Press: Boulder, 2000; pp 102-137.

61. Chase, A. F.; Chase, D. Z.; White, C. In *Reconstruyendo la Ciudad Maya: El Urbanismo en las Sociedades Antiguas*; Ciudad Ruiz, A.; Ponce de Leon, M. J. I.; Martinez Martinez, M. D. C., Eds.; Sociedad Espanola de Estudios Mayas: Madrid, Spain, 2001; pp 95-122.
62. White, C. D.; Healy, P. F.; Schwarcz, H. P. *Journal of Anthropological Research* **1993**, *49*, 347-375.
63. Whittington, S. L.; Reed, D. M. In *Bones of the Maya: Studies of Ancient Skeletons*; Whittington, S. L.; Reed, D. M., Eds.; Smithsonian Institution Press: Washington, DC, 1997; pp 157-170.
64. Reed, D. M. PhD dissertation, The Pennsylvania State University, State College, PA, 1998.
65. Whittington, S. L.; Reed, D. M. In *VII Simposio de Investigacions Arqueológicas en Guatemala, 1993*; Laporte, J. P.; Escobedo, H. L., Eds.; Museo Nacional de Arqueología y Etnología: Guatemala, 1994; pp 23-28.
66. Whittington, S. L.; Reed, D. M. *Mesoamérica* **1998**, *35*, 73-82.
67. Wright, L. E.; Schwarcz, H. P. *American Journal of Physical Anthropology* **1998**, *106*, 1-18.
68. Wright, L. E.; Schwarcz, H. P. *Journal of Archaeological Science* **1999**, *26*, 1159-1170.
69. Lee-Thorp, J. A.; Sealy, J. C.; van der Merwe, N. J. *Journal of Archaeological Science* **1989**, *16*, 585-99.
70. Coyston, S.; White, C. D.; Schwarcz, H. P. In *Reconstructing Ancient Maya Diet*; White, C.D., Ed.; University of Utah Press: Salt Lake City, 1999; pp 221-244.
71. Whittington, S. L.; Tykot, R. H.; Reed, D. M. *Ancient Mesoamerica*. Submitted.
72. Hodell, D. A.; Curtis, J.H.; Brenner, M. *Nature* **1995**, *375(6530)*, 391-394.
73. Leyden, B. W.; Brenner, M.; Dahilin, B. *Quaternary Research* **1998**, *49*, 111-122.
74. Wright, L. E. In *Bones of the Maya: Studies of Ancient Skeletons*; Whittington, S. L.; Reed, D. M., Eds.; Smithsonian Institution Press: Washington, DC, 1997; pp 181-195.
75. White, C. D. In *Bones of the Maya: Studies of Ancient Skeletons*; Whittington, S. L.; Reed, D. M., Eds.; Smithsonian Institution Press: Washington, DC, 1997; pp 171-180.
76. White, C. D.; Wright, L. E.; Pendergast, D. M. In *In the Wake of Contact. Biological Responses to Conquest*; Larsen, C. S.; Milner, G. R., Eds.; Wiley-Liss: New York, 1994; pp 135-145.
77. Danforth, M. E. In *Reconstructing Ancient Maya Diet*; White, C.D., Ed.; University of Utah Press: Salt Lake City, 1999; pp 103-118.
78. Storey, R. In *Reconstructing Ancient Maya Diet*; White, C.D., Ed.; University of Utah Press: Salt Lake City, 1999; pp 169-182.

## Chapter 15

# Chemical Compositions of Chinese Coins of Emperor Ch'ien Lung (Qian Long) and Annamese Coins of Emperor Thanh Thai via Energy-Dispersive X-ray Fluorescence

**Tiffany Gaines, Eric McGrath, Vincent Iduma, Renee Kuzava, Sarah Frederick, and Mark Benvenuto**

**Department of Chemistry and Biochemistry, University of Detroit Mercy, 4001 West McNichols Road, Detroit, MI 48219-0900**

Two hundred twenty five Chinese and one hundred Annamese coins were analyzed via energy dispersive X-ray fluorescence spectrometry for the following elements: copper, zinc, tin, lead, arsenic, and antimony. The composition of both series of coins varied significantly in the amount of lead present in various samples. The Annamese coins, though cast roughly one century after the Chinese, still had roughly similar levels of trace elements (possibly impurities) present. The Annamese series and the Chinese coins of the southern branch mints did not contain correspondingly higher levels of antimony, implying restricted trade between the two regions.

Casting coins from brass, or some brass-like alloy has a long history in China, Korea, Japan, and southeast Asia, one that reaches back into ancient times. The traditional coin is referred to as a "cash", a round coin with a square, central hole, and cash were usually strung together for ease of transport and trade. There has been a certain amount of numismatic study of such coins, with the emphasis being on those of China, as opposed to the coins of Korea, Japan, or what is now Vietnam (*1-3*). There appear to have been no metallurgical studies of these coins however, using either destructive or non-destructive techniques, perhaps because their low cost and value makes them less desirable to study as series of artifacts.

The Chinese series of coins in this study all are attributed to the emperor Ch'ien Lung (in Pinyin transliteration, Qian Long) who reigned from 1736-1796 (*1*). The two sides of the coin are designated the obverse, which shows four characters, and the reverse, which shows two characters in a different script than that used on the obverse. The obverse is in a Chinese script read by most people living in China, while the reverse is a Manchu script not normally read by the majority of people in China today (*3*). The obverse characters are read: top, bottom, right, left, and invariably translate as: "the money of Ch'ien Lung." The reverse two characters indicate the mint city in which the coin was cast.

The Annamese series of coins are attributed to the emperor Thanh Thai, who reigned from 1880-1907 (*4*). The obverse shows four characters, in the same manner as the Chinese coins, including a similar arrangement, while the reverse displays only one character, indicating its value. Although the Annamese coins are of a more recent time frame, the casting technology for them is the same as that for the Chinese series of coins.

These two series of coins, though differing in time of production by roughly a century, are comparable precisely because of Annam's traditional, historical, and cultural relationship to imperial China. Since at least the time of the voyages of discovery by the imperial Chinese navy in the early 1400s (*5*), Chinese culture, art, and technology have been imitated and adapted by Annamese craftsmen, as well as those in other eastern Asian lands. At times throughout history, Annam has been essentially a province of greater China. While elements within the Japanese imperial government chose to modernize and westernize their operations in the nineteenth century, in China and several other lands influenced by Chinese culture (Annam included) the decision was made to continue many governmental functions and operations in the traditional manner. This included the ancient methods of casting coins, as well as the traditional methods of metal refining.

Energy dispersive X-ray fluorescence spectrometry was applied to a large sample number (225 Chinese and 100 Annamese coins) in order to determine what trends could be discerned in chemical composition. Though all of the coins are supposed to be brass, and though brass objects, including coins, from various cultures have been studied extensively, (*6-22*) several elements besides copper and zinc are present in these two series of objects.

## Procedure

Each of the coins in both series was prepared only by wiping with a dry Kimwipe or a dry toothbrush to remove any loose soil or encrustation. In addition, coins were visually inspected to ensure that a clean surface could be exposed to the X-ray beam. No further physical or chemical modification of the coins' surfaces was performed. After this cleaning and inspection, the coins were examined for Cu, Zn, Sn, Pb, Fe, Ni, As, Sb, Au, Pt, Pd, and Ag via energy dispersive X-ray fluorescence on a Kevex Spectrace Quanx instrument. Sample excitation conditions were as follows: 20 kV, 0.10 mA, 100 second count, $K\alpha,\beta$ for Fe, Ni, Cu, Zn, As, Pt, Au, and Pb, followed by 45 kV, 0.72 mA, 60 second count, L lines, for Pd, Ag, Sn, and Sb, using a rhodium target X-ray tube. Fundamental parameters software and pure element standards were utilized in determining major and minor concentrations. Brass standards of known concentrations were run after each 100 samples, as a minimum.

The Chinese coins, having been cast at several different mints, were grouped according to the minting cities (*1*) and graphed against elemental compositions, while the Annamese coins were examined by comparing number of coins against percentage of various elements (since there appears to be no reference book or paper delineating casting furnace locations for them). The numbering system for the Chinese series utilizes the Schjoth reference numbers (*1*), and is listed here with a pinyin transliteration. As well, the number of coins and general region of each mint is listed.

| | No. of coins | Region |
|---|---|---|
| - 1464, Beijing, Board of Revenue | 25 | north |
| - 1465, Beijing, Board of Revenue | 67 | north |
| - 1466, Beijing, Board of Works | 49 | north |
| - 1467, Chilhi provincial mint | 4 | central |
| - 1468, Shaansi provincial mint | 3 | central |
| - 1469, Fujian provincial mint | 2 | southeast |
| - 1470, Zhejiang provincial mint | 8 | east |
| - 1471, Jiangsu provincial mint | 6 | east |
| - 1475, Nanchang provincial mint | 3 | central |

| | | | |
|---|---|---|---|
| - | 1476, Guilin provincial mint | 19 | southeast |
| - | 1478, Fujian provincial mint | 6 | southeast |
| - | 1480, Yunnan provincial mint | 23 | south |
| - | 1483, Hunan provincial mint | 2 | south |
| - | 1484, Sichuan provincial mint | 5 | south |
| - | 1485, Shanxi provincial mint. | 3 | north |

## Chemical Compositions

The chemical compositions of the Chinese coins of Emperor Ch'ien Lung are expressed graphically in Figures 1–6b, while the Annamese coins of Thanh Thai are shown in Figures 7-9. It can be seen that in most cases copper was indeed the chief component of these samples, though the percentages vary widely, with a few samples in each series dipping below 50%, as shown in Figures 1 and 7. As well, percentages of zinc vary considerably, with a few samples in each series containing over 50% zinc, as seen in Figures 2 and 7. If the coins in either series had been undiluted brass, the graphs of copper versus zinc percentages would be ratios in the range of 1:1 to 3:1; this is certainly not the case for many of the samples (the Chinese series ratios are represented in Figure 4). Within the Chinese series, it is evident that higher zinc concentrations do occur in several of the southern branch mints, such as Guilin, Fujian, and Yunnan. These are however only a small subset within the entire series. Likewise, the Annamese series contains a few pieces of relatively high zinc content, as seen in Figure 7. The reason for these high zinc percentages (when zinc in standard, modern brass is seldom as high as 50%) is not immediately obvious, but may relate to the cost of materials or the look of the finished coins. Zinc routinely costs less than copper, and these high zinc examples still have the color of brass.

The coins were also analyzed for tin, since it has been used throughout history in the metallurgy of various cultures rather interchangeably with zinc, and since bronze has been used in lieu of brass for many objects. Tin concentrations were found to be uniformly low however, with the highest concentrations (in two samples, 10% and 11%) occurring in only a few coins of the Annamese series, shown in Figure 7.

Lead seems to be an element that has been added almost at random, with percentages ranging from a fraction of 1% to as high as 26%, as shown in Figures 3 and 7. Numismatic sources indicate that foundry masters were paid based on the number of coins they produced, and therefore had a financial incentive to debase their alloys and make more coins (*3*). As well, in the Chinese series, one source indicates that 6.5% lead was allowed as part of the

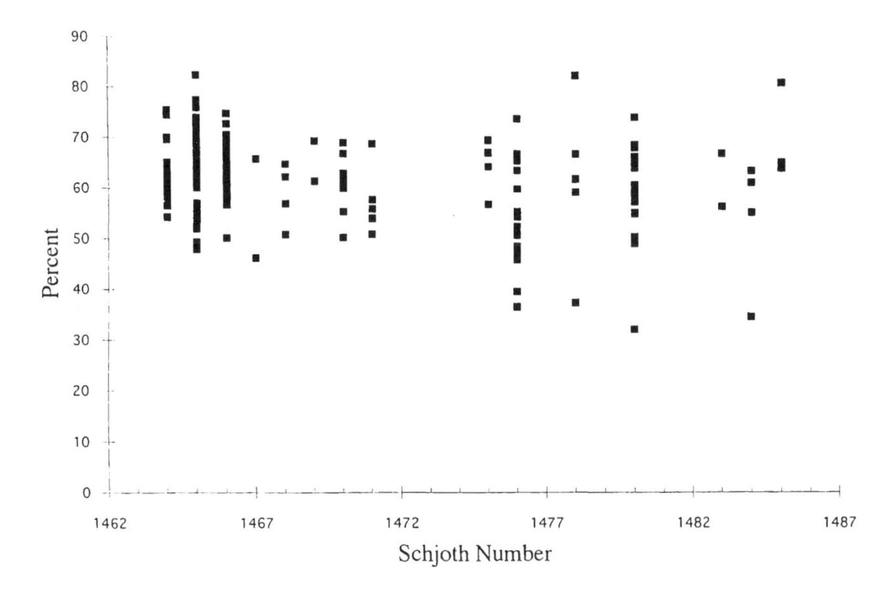


Figure 1. Copper percent, Chinese.

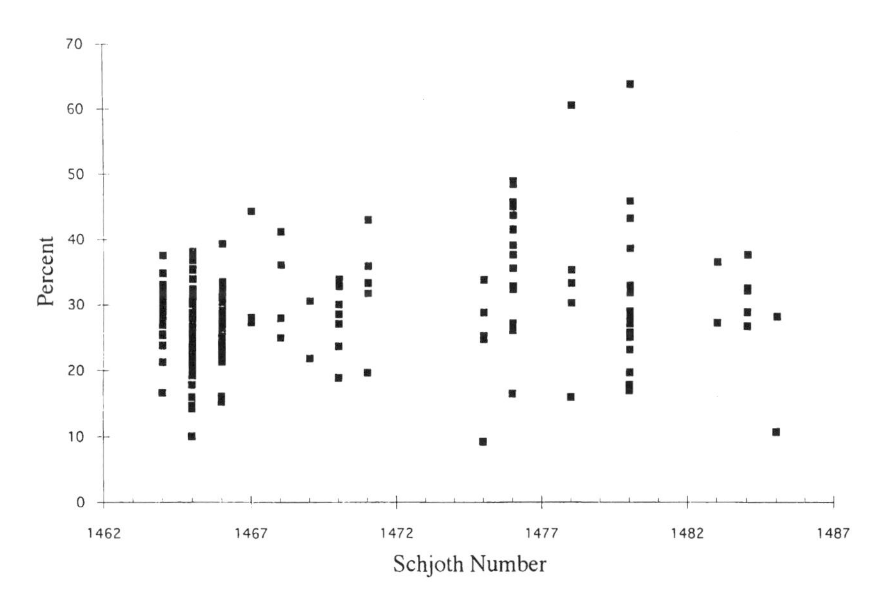


Figure 2. Zinc percent, Chinese.

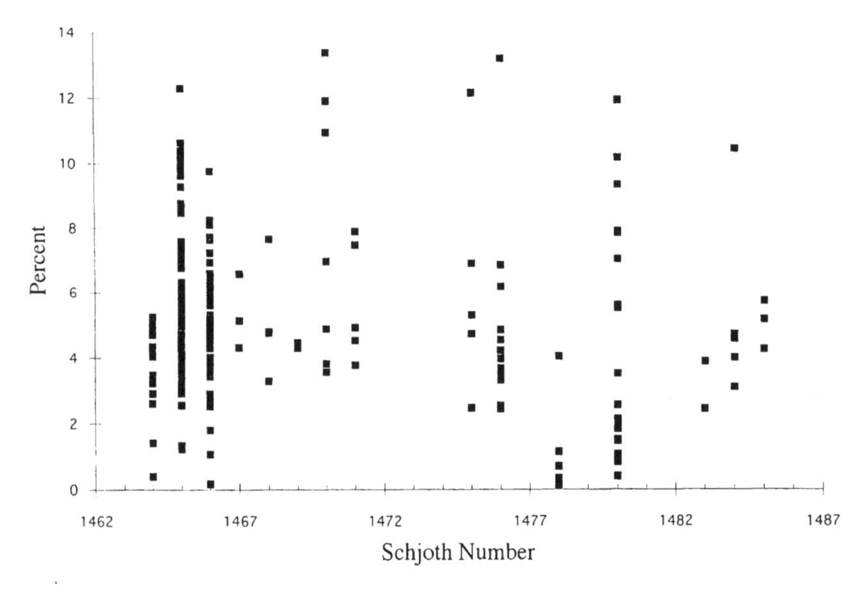


Figure 3. Lead percent, Chinese.

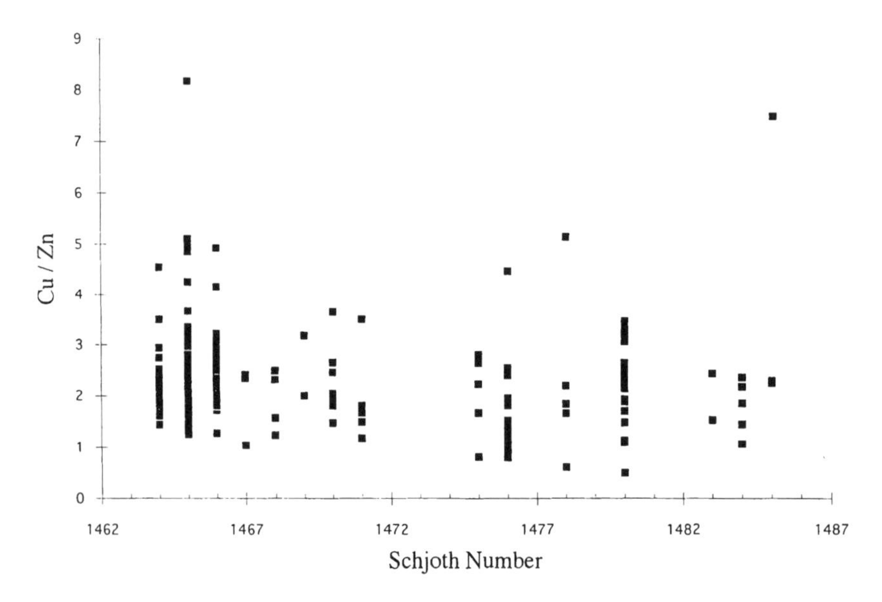


Figure 4. Cu/Zn ratio, Chinese.

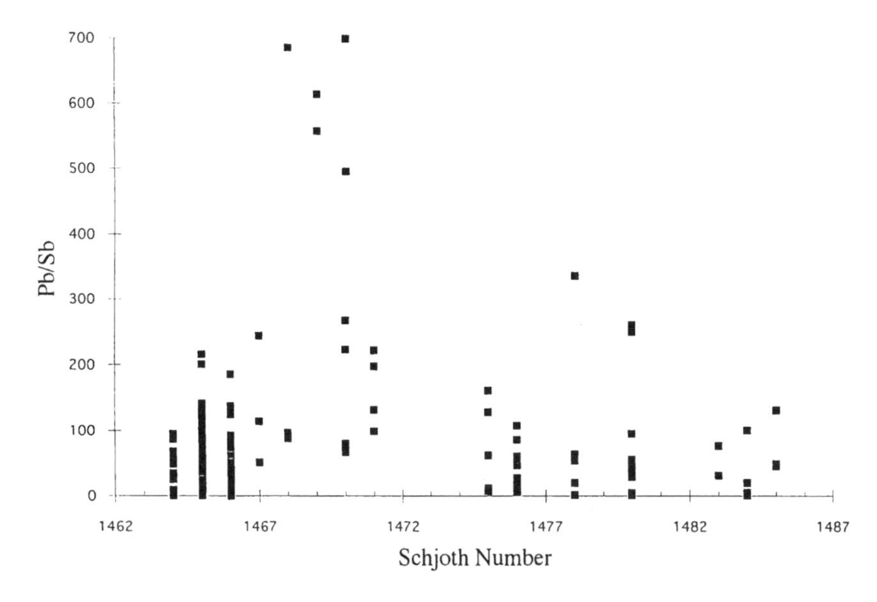


Figure 5a. Pb/Sb ratio, Chinese.

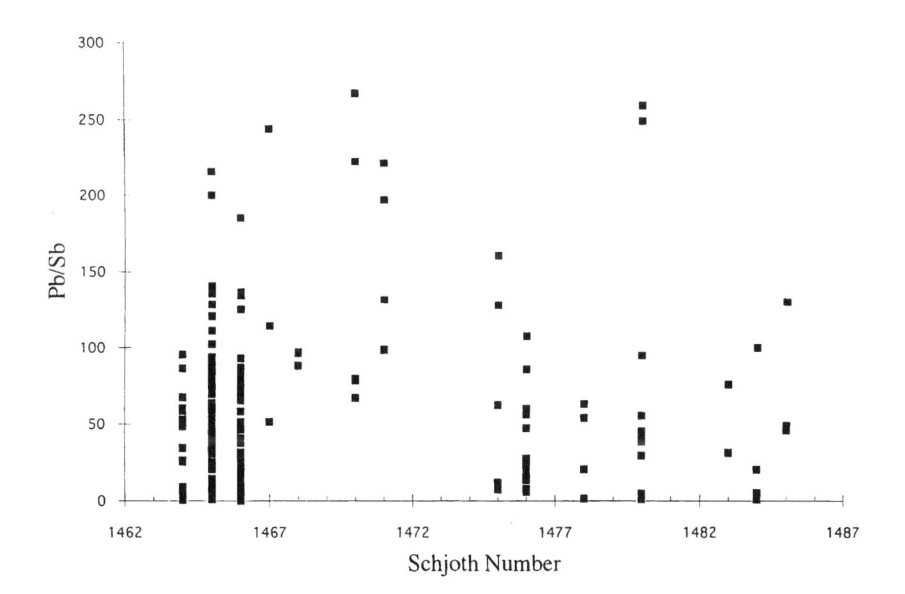


Figure 5b. Pb/Sn ratio, Chinese.

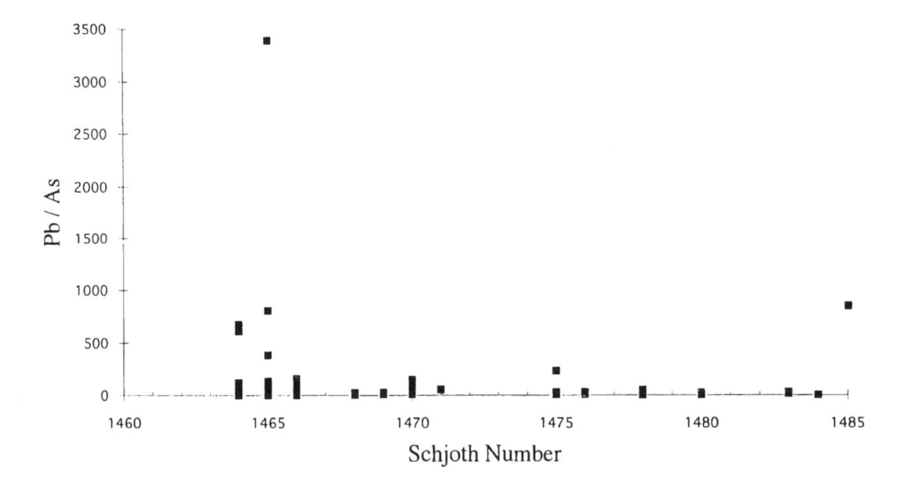


Figure 6a. Pb/As ratio, Chinese.

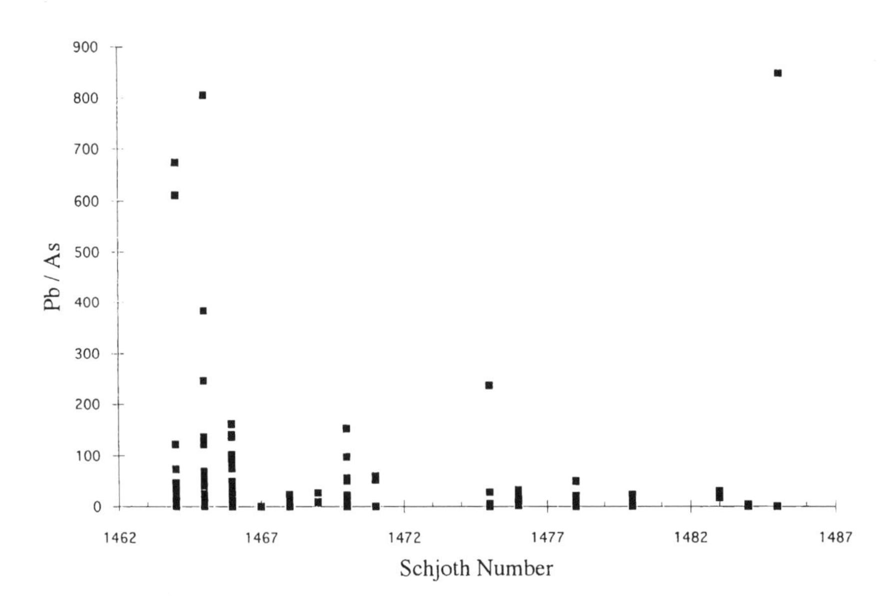


Figure 6b. Pb/As ratio, Chinese.

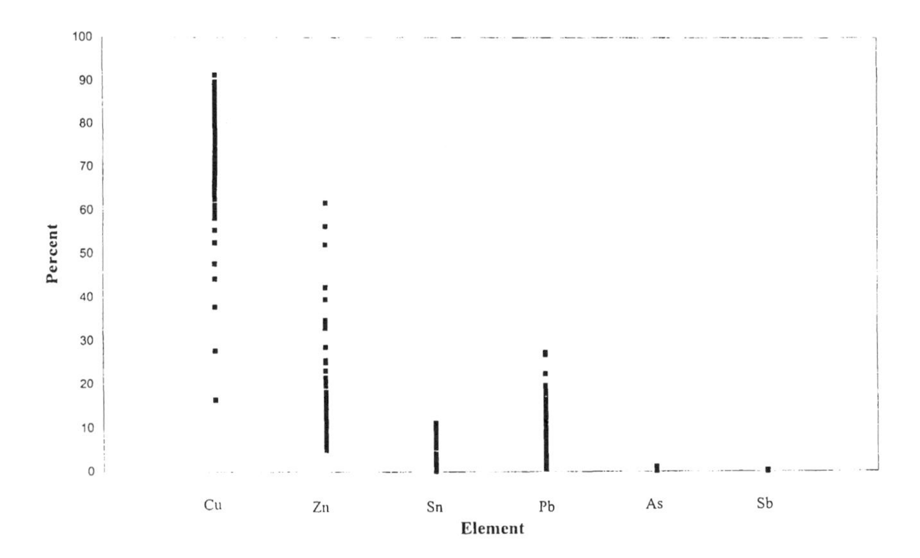


Figure 7. Cu, Zn, Sn, Pb, As, Sb percentages, Annamese.

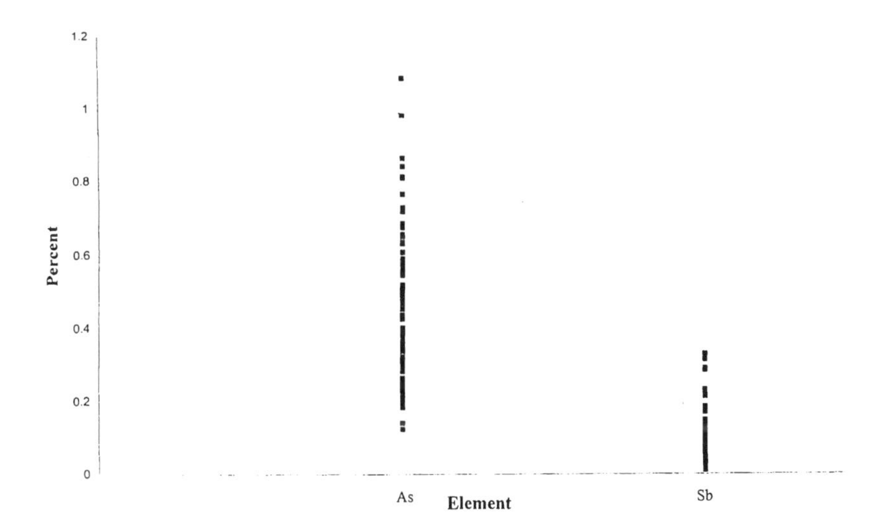


Figure 8. As, Sb percentages, Annamese.

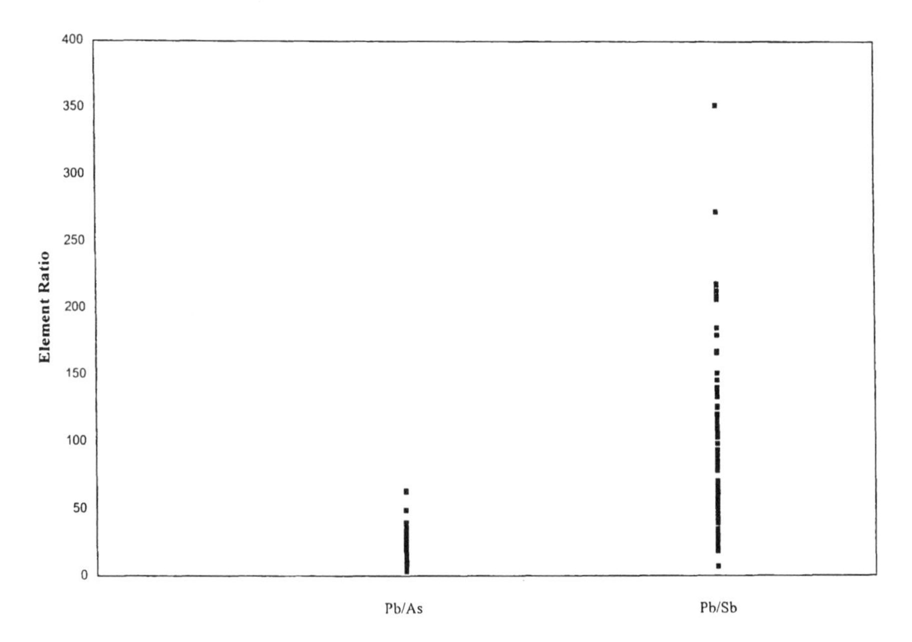


Figure 9. Pb/As, Pb/Sb ratios, Annamese.

official alloy (*3*); and lead concentrations in coins from the three Beijing foundries do appear to have a preponderance of samples near that concentration, though the outliers in those subsets vary from trace lead to over 10%.

Lead concentrations were examined in relation to arsenic and antimony, because of their mutual occurrence in various ores (*23*), which is implied by their proximate positions on the periodic table. Ratios of lead to these elements were constructed, as shown in Figures 5, 6 and 9, in an attempt to see if patterns exist. High ratio numbers indicate lead (which may be high or low itself) with only traces of the lesser element, while low ratio values might prove themselves valuable in determining a source of a lead ore, again assuming lead and the lesser element were mined together. In the Chinese series, it appears that the Beijing foundries (Schjoth numbers 1464-1466) contained the lowest Pb/trace-element ratios. Therefore Figures 5 and 6 were re-graphed as Figures 5b and 6b without the highest ratio points, to make the more densely packed low ratio data points more clearly discernible. The Annamese series shows a wide range of ratios in Figure 9 (especially the Pb/Sb ratios), which at first appears confusing, but which may simply indicate a large number of different ore sources, specifically lead sources. Unfortunately, these Pb/trace-element ratios do not reveal an obvious pattern that indicates ore sources, which implies that antimony and arsenic may have been present in lead ores as well as other elemental ore sources.

Overall, it appears that the coin chemistry displayed in the coins of individual Chinese mints do not differ widely. Rather, they all show similar compositional patterns.

While the coins in each series were examined for iron, nickel, gold, platinum, palladium, and silver, these six elements were found only in trace amounts, and appear to have no direct correlation to copper, zinc, tin, or even lead. Initial assumptions were that these six elements might provide a type of chemical fingerprint that could lead to one or more ore sites. Unfortunately, no such connection was discernible.

In a similar vein, it was hoped that a correlation would be found between higher antimony levels and the Annamese coins, as well as the Chinese coins of the southern mints, especially Hunan, since this province is a major source of antimony today. The presence of antimony could then be used as a chemical fingerprint for the ores from which the coins' metals were produced. That correlation does not appear to be readily apparent however. In the Chinese series, less than five coins have higher than 5% antimony concentration, while in the Annamese series, the antimony concentration only rises above 1% once. In like fashion, arsenic occurs as only a trace component of the coins.

Our similar study of Japanese coins of the same general time, the mid-1800s (*24*), found them to be essentially high copper, leaded bronze, with consistently low concentrations of tin and trace concentrations of zinc. This is a large difference in composition when compared to these two series, and appears, as mentioned in the introduction, to reflect the known trend towards modernization Japan experienced in the nineteenth century, and the resistance to that trend that was seen in China and Annam as well.

## Conclusions

The Chinese and Annamese coins studied appear in all cases to be brass, but of impure compositions that could be described as "leaded, bronzy, brass" because of the presence of varying amounts of lead and tin. In addition, elements such as arsenic and antimony are found in almost all the coins, possibly indicating that the quality control in ore refining had declined from previous times (*9*), or indicating some unscrupulousness on the part of the mint masters (*3*).

The presence of heightened antimony levels in certain of the Chinese coins from southern mints, and the corresponding lack thereof in the Annamese coins, may reflect a restriction or lack of trade between the two, at least during the time of the casting of the Annamese pieces, in the late nineteenth century.

## Acknowledgements

The authors wish to thank the United States National Science Foundation, grant number 98-51311 as well as the University of Detroit Mercy for the purchase of the energy dispersive X-ray fluorescence spectrometer, and Project SEED for V.I.'s involvement in this research.

## Literature Cited

1. Schjoth, F. *Chinese Currency;* University of Oslo: Oslo, Norway, 1929; pp 64-66.
2. Lockhart, J.H.S. *The Lockhart Collection of Chinese Copper Coins*; Quarterman Publications, Inc.: Lawrence, MA, 1975.
3. Hartill, D. *Numismatic Chronicle* **1991**, *151*, 67-120.
4. *Standard Catalog of World Coins*; Mishler, C., Eds.; Krause Publications, Inc.: Iola, WI, 1998.
5. Boorstin, D.J. *The Discoverers: A History of Man's Search To Know His World and Himself;* Vintage Books, Random House: New York, NY, 1985.

6. Meyers, P.; Van Zelst, L.; Sayre, E.V. In *Archaeological Chemistry;* Beck, C.W., Ed.; Advances In Chemistry Series 138; American Chemical Society: Washington, DC, 1974; pp. 22-33.
7. Chase, W.T. In *Archaeological Chemistry;* Beck, C.W., Ed.; Advances In Chemistry Series 138; American Chemical Society: Washington, DC, 1974; pp. 148-185.
8. Goucher, C.L.; Teilhet, J.H.; Wilson, K. R.; Chow, T.J. In *Archaeological Chemistry II;* Carter, G.F., Ed.; Advances In Chemistry Series 171; American Chemical Society: Washington, DC, 1978; pp. 278-292.
9. Chase, W.T.; Ziebold, T.O. In *Archaeological Chemistry II;* Carter, G.F., Ed.; Advances In Chemistry Series 171; American Chemical Society: Washington, DC, 1978; pp. 293-334.
10. Goad, S.I.; Noakes, J. In *Archaeological Chemistry II;* Carter, G.F., Ed.; Advances In Chemistry Series 171; American Chemical Society: Washington, DC, 1978; pp. 335-346.
11. Carter, G.F. In *Archaeological Chemistry II;* G.F. Carter, Ed.; Advances In Chemistry Series 171; American Chemical Society: Washington, DC, 1978; pp. 347-377.
12. Rapp, Jr. G.; Allert, J. In *Archaeological Chemistry III;* Lambert, J.B., Ed.; Advances In Chemistry Series 205; American Chemical Society: Washington, DC, 1984; pp. 273-293.
13. Carter, G.; In *Archaeological Chemistry III;* Lambert, J.B., Ed.; Advances In Chemistry Series 205; American Chemical Society: Washington, DC, 1984; pp. 311-329.
14. Gale, N.H.; Stos-Gale, Z.A. In *Archaeological Chemistry IV;* Allen, R.O. Ed.; Advances In Chemistry Series 220; American Chemical Society: Washington, DC, 1989: pp. 159-198.
15. Carter, G.F.; Razi, H. In *Archaeological Chemistry IV;* Allen, R.O. Ed.; Advances In Chemistry Series 220; American Chemical Society: Washington, DC, 1989: pp. 213-230.
16. Hopke, P.K.; Williams, W.S.; Maguire, H.; Farris, D. In *Archaeological Chemistry IV;* Allen, R.O. Ed.; Advances In Chemistry Series 220; American Chemical Society: Washington, DC, 1989: pp. 233-247.
17. Moreau, J-F.; Hancock, R.G.V. In *Archaeological Chemistry: Organic, Inorganic, and Biochemical Analysis;* Orna, M.V., Ed.; Advances in Chemistry Series 625; American Chemical Society: Washington, DC, 1996; pp. 64-82.
18. Carter, G.F. In *Archaeological Chemistry: Organic, Inorganic, and Biochemical Analysis;* Orna, M.V., Ed.; Advances in Chemistry Series 625; American Chemical Society: Washington, DC, 1996; pp. 94-106.

19. Butcher, K.; Ponting, M. *American Journal of Numismatics* **1997,** *9,* 17-36.
20. Carter, G.F. *American Journal of Numismatics* **1997,** *7-8,* 235-250.
21. Mabuchi, H.; Notsu, K.; Nishimatsu, S.; Fuwa, K.; Iyama, H.; Tominaga, T. *Nippon Kagaku Kaishi* **1979,** *5,* 586-90.
22. Lutz, J.; Pernicka, E. *Archaeometry* **1996,** *38,* 313-323.
23. Greenwood, N.N.; Earnshaw, A. *Chemistry of the Elements;* Pergamon Press: Oxford, England, 1984; pp 638-640.
24. Gaines, T.; McGrath, E.; Langrill, R.; Benvenuto, M. *Numismatic Chronicle* **2001,** *161,* 308-317.

# INDEXES

# Author Index

# Subject Index

## C

**D**



**E**

## F

## G

**M**

**N**

## S

**T**